Self-Sufficient Agriculture

Labour and Knowledge in Small-Scale Farming

Robert Tripp

With Catherine Longley, Caleb Mango,
Nelson Mango, Wilson Nindo,
V. Hiroshini Piyadasa, Michael Richards,
Laura Suazo and Mahinda Wijeratne

London • Sterling, VA

First published by Earthscan in the UK and USA in 2006

Copyright © ODI, 2006

ISBN-13:   978-1-84407-297-2   paperback
ISBN-10:   1-84407-297-5   paperback
ISBN-13:   978-1-84407-296-5   hardback
ISBN-10:   1-84407-296-7   hardback

Typesetting by FISH Books, Enfield, Middlesex
Printed and bound in the UK by Cromwell Press, Trowbridge
Cover design by Danny Gillespie

For a full list of publications please contact:

Earthscan
8–12 Camden High Street
London, NW1 0JH, UK
Tel: +44 (0)20 7387 8558
Fax: +44 (0)20 7387 8998
Email: earthinfo@earthscan.co.uk
Web: **www.earthscan.co.uk**

22883 Quicksilver Drive, Sterling, VA 20166-2012, USA

Earthscan is an imprint of James and James (Science Publishers) Ltd and publishes in association with the International Institute for Environment and Development

A catalogue record for this book is available from the British Library

Library of Congress Cataloging-in-Publication Data

Tripp, Robert.
   Self-sufficient agriculture: labour and knowledge in small-scale farming / Robert Tripp.
      p. cm.
   Includes bibliographical references and index.
   ISBN 1-84407-279-5 – ISBN 1-84407-296-7
   1. Agricultural innovations–Developing countries. 2. Sustainable agriculture–Developing countries. 3. Farms, Small–Developing countries. I. Ttile.
   S494.5 I5T75 2005
   630'.9172'4–dc22   200518171

Contents

List of Figures, Tables and Boxes   *viii*
List of Contributors   *x*
Preface   *xi*
List of Acronyms and Abbreviations   *xiii*

1   Low External-Input Technology (LEIT) and Agricultural
    Development   1
    Robert Tripp

    External inputs and agricultural development   1
    Sustainable agriculture   3
    Low external-input technology (LEIT)   10
    Issues related to the performance of LEIT   13
    The nature of the analysis   17

2   Examples of LEIT and Farmer-Focused Development Strategies   21
    Robert Tripp

    Introduction   21
    Soil and water management   23
    Soil fertility enhancement   27
    Crop establishment   31
    Controlling weeds, insects and pathogens   33
    Methods for generating and promoting LEIT   36
    Summary   41

3   Labour, Information and Agricultural Technology   43
    Robert Tripp

    Introduction   43
    Labour   45
    Information   52
    Human capital   55
    Social capital   59
    Summary   66

4   The Impact of LEIT: Evidence from the Literature   68
    Robert Tripp

    Introduction   68
    The poverty focus of low external-input technology (LEIT) 69
    Economic incentives   74
    Labour   78
    Human capital   83
    Political movements in support of LEIT   88
    Social capital and information flow   90
    Summary   93

5   Learning from Success: Revisiting Experiences of LEIT Adoption
    by Hillside Farmers in Central Honduras   95
    Michael Richards and Laura Suazo

    Introduction   95
    Methodology   98
    General results   101
    Why do some farmers adopt and others not?   107
    Innovative farmers   113
    Dissemination pathways and information flows   113
    Project impacts on social and human capital   114
    Were the 'low external-input technology' (LEIT)
    adopters 'low-input' farmers?   115
    Summary   116
    Annex 1: Logit econometric analysis   118
    Annex 2: Multiple regression analysis of basic grain yields 120
    Annex 3: Results of the crop budgeting exercise   122

6   Conservation by Committee: The Catchment Approach to Soil
    and Water Conservation in Nyanza Province, Western Kenya 125
    Catherine Longley, Nelson Mango, Wilson Nindo and Caleb Mango

    Introduction   125
    The evolution of soil and water conservation in Kenya   126
    Background to the National Soil and Water Conservation
    Programme (NSWCP)   127
    Background to the case study area and the fieldwork sites 131
    Who participated in the Catchment Approach?   137
    Uptake of the different technologies   138
    Levels of technology adoption   142
    Farmers' sources of information about soil and water conservation
    (SWC)   148
    Experimentation, modifications and innovation   149
    Enhancing human capital: Increased knowledge   150
    Enhancing social capital: Local groups and organizations 151
    Summary   153
    Annex 1: Methodology   155

7   After School: The Outcome of Farmer Field Schools in
    Southern Sri Lanka   160
    Robert Tripp, Mahinda Wijeratne and V. Hiroshini Piyadasa

    Introduction   160
    Rice, extension, and integrated pest management (IPM)   160
    Research area and organization of the study   162
    Who participates in farmer field schools (FFS)?   166
    The effects of the FFS   169
    The diffusion of knowledge   175
    Factors related to the use of new production practices   179
    Summary   183

8   The Trajectory of Low External-Input Agriculture   187
    Robert Tripp

    Introduction   187
    The utilization of low external-input technology (LEIT)   188
    Participation in LEIT projects   190
    The type of farmer who utilizes LEIT   191
    Labour and LEIT   194
    Knowledge and LEIT   196
    The emergence of a 'LEIT farmer'   198
    LEIT, experimentation and adaptation   200
    Farmer-to-farmer dissemination of LEIT   201
    LEIT and social capital   202
    Summary of case study experience   203
    Technology and agricultural development   204
    Assessing LEIT   206
    Projects and scaling up   209
    The way forward   213

References   *219*
Index   *235*

List of Figures, Tables and Boxes

Figures

5.1   Study communities in Franciso Morazán Department, Central
      Honduras   97
5.2   Adoption of technologies by region, excluding control farmers   105
5.3   Tendencies in the adoption and abandonment of in-row tillage
      (IRT) in the three regions   106
5.4   Tendencies in the adoption and abandonment of green manures
      (GRMs) in the three regions   106
6.1   Location of study sites in Nyanza Province, western Kenya   136
6.2   Implementing soil and water conservation (SWC) structures
      over time   141
7.1   Locations of study sites in Sri Lanka   163
7.2   Relationship between insecticide use and soil fertility practices
      (all farmers)   181
8.1   Relation between low external-input technology (LEIT)
      use and other low-input practices   199
8.2   Examples from Honduras   199
8.3   Examples from Kenya   199
8.4   Examples from Sri Lanka   199

Tables

5.1   Communities studied and sample size   99
5.2   Distribution of participants, non-participants and control farmers   100
5.3   General farming characteristics   102
5.4   Comparisons between participants and
      non-participants (regions 1–2)   103
5.5   Comparison between leaders and normal participants (regions 1–2) 103
5.6   Adoption of project technologies (percentage of all farmers)   104
5.7   Adoption of low external-input technologies (LEITs) by participant
      farmers (including leaders)   105
5.8   Comparison of adopter and non-adopter farmers (regions 1–2)   107
5.9   Characteristics of farmers who never tried, abandoned and
      continue using in-row tillage (IRT) (project communities
      and participants in regions 1–2)   108
5.10 Adoption according to resource wealth category (regions 1–2)   109

5.11   Summary of social capital indices (regions 1–2)   112
5.12   Use of chemical and organic inputs with/without LEITs
       (regions 1–2)   116
A5.1   Odds ratios in the three regressions (standard errors)   119
A5.2   Marginal effects at sample median values of regressors   120
A5.3   Average yields of basic grains (regions 1–2)   120
A5.4   Partial budget analysis of main technology–crop associations
       (regions 1–2)   123
6.1   Key features of the study sites   133
6.2   Three most important sources of household income during
       the past year   134
6.3   Sample farmers   136
6.4   Number of farms in sample catchments   137
6.5   Characteristics of committee members compared with
       other catchment farmers   138
6.6a   Period of implementation of soil and water conservation
       (SWC) structures in high-potential sites   139
6.6b   Period of implementation of SWC structures in low-potential sites   140
6.7a   Number and density of grass/unploughed strips in
       high-potential sites   144
6.7b   Number and density of grass/unploughed strips in
       low-potential sites   145
6.8a   Levels of adoption in high-potential sites (percentage of
       each farmer type)   146
6.8b   Levels of adoption in low-potential site (percentage of each
       farmer type)   146
6.9   Factors relating to the adoption of SWC technologies   147
6.10   Adoption levels pre- and post-project   147
7.1   Research locations   165
7.2   Farmer field school (FFS) farmers and their neighbours   167
7.3   Insect-control practices of farmer field school (FFS) farmers
       and their neighbours   170
7.4   Farmers' knowledge and opinions on insect control   172
7.5   Soil fertility practices of farmer field school (FFS) farmers and
       their neighbours   173
7.6   Farmers' sources of ideas for lowering insecticide use
       (percentage of farmers)   177
7.7   Most frequent instances of information transmission from
       farmer field school (FFS) farmers to their neighbours (n = 70)   178
7.8   Relationship of beliefs and knowledge to insecticide use   180
7.9   Principal occupations and cultivation practices   183

Boxes

6.1   Farmers' levels of adoption   143
6.2   An example of farmer-to-farmer learning   149

List of Contributors

Catherine Longley is a research fellow in the Overseas Development Institute's (ODI's) Humanitarian Policy Group based in Nairobi, working in collaboration with the International Crops Research Institute for the Semi-Arid Tropics (ICRISAT).

Caleb Mango is a natural resource manager with Earthview Geoconsultants International, Nairobi.

Nelson Mango is a social scientist with the International Livestock Research Institute (ILRI), Nairobi.

Wilson Nindo is a facilitator for community-based research and training and a consultant with the World Agroforestry Centre in western Kenya.

V. Hiroshini Piyadasa is an agriculturalist currently pursuing postgraduate studies at the University of Peradeniya, Sri Lanka.

Michael Richards is an agricultural economist working at the International NGO Training and Research Centre (INTRAC) and was formerly with the Overseas Development Institute (ODI).

Laura Suazo is an agronomist and agricultural education specialist currently working as a research associate with the International Centre for Cover Crop Information (CIDICCO), Tegucigalpa.

Robert Tripp is a social anthropologist and a research fellow in the Rural Policy and Governance group at the Overseas Development Institute (ODI).

Mahinda Wijeratne is an agricultural economist and professor in the Faculty of Agriculture at the University of Ruhuna, Sri Lanka.

Preface

It is not difficult to understand why low external-input agriculture has attracted so much interest and support in discussions about the future of farming in developing countries. Continuing rural poverty, the high cost of purchased inputs and a growing list of environmental problems all support the view that farmers should rely as much as possible upon local resources to enhance their soils and manage their crops. Although a precise and widely accepted definition of sustainable agriculture eludes us, the substitution of local resources for purchased inputs is certainly a common theme. For some, the goal is a pragmatic one that seeks the judicious combination of local resources and external inputs. For others, a complete rejection of conventional agricultural technology is seen as the answer to smallholder development, providing a coherent prescription for ensuring rural dignity and prosperity.

This book does not examine the wider controversies associated with various visions of sustainable agriculture, but rather focuses on the set of crop management alternatives, described here as low external-input technology (LEIT). Because LEIT is so often caught up in more general discussions about the direction of agricultural development, it does not receive the objective assessment that it deserves. On the one hand, LEIT is often promoted as not only the basis of self-sufficiency and environmentally friendly agriculture, but also as a key to the development of human and social capital in farming communities. On the other hand, critics often portray LEIT as a set of hopelessly labour intensive and awkward techniques that offer few contributions for the development of modern farming. However, the literature provides surprisingly few careful assessments of the performance of specific instances of LEIT. The problem is compounded by the fact that most descriptions of LEIT are concerned with ongoing or recently completed projects and do not allow us to follow the trajectory of LEIT in farmers' hands.

Discussions with Michael Richards and Catherine Longley at the Overseas Development Institute (ODI) regarding weaknesses in the assessment of LEIT led to the development of a research proposal that focused on examining the outcome of past projects that had been successful at promoting LEIT. We sought to identify examples where the introduction of LEIT had been part of well-managed projects (to avoid focusing on administrative issues) and covered a relatively wide area (to avoid pilot project experiences). We believed that we could learn a great deal about the nature of LEIT by asking what was happening five years after the completion of such projects. To what extent had farmers continued to use the innovations that had been introduced? Were there clear differences between those farmers who were able to take advantage of the technology and those who did not? To what extent did farmers further refine and adapt the technologies? Was there evidence of farmer-to-farmer diffusion of these ideas and techniques? Did the experience of successfully participating in a LEIT project lead to further growth in human or social capital? It proved more difficult than expected to identify suitable cases; but we eventually found three excellent examples and pursued field research in collaboration with partners in Honduras (Laura Suazo), Kenya (Nelson Mango, Wilson Nindo and Caleb Mango) and Sri Lanka (Mahinda Wijeratne and V. Hiroshini Piyadasa).

A core set of questions formed the basis for field research, as well as guiding a thorough review of the literature. Given the wide range of environments and technologies included in the study, the answers to many of the basic questions are surprisingly consistent. They provide support for the argument that many examples of LEIT offer significant contributions to improving smallholder productivity, livelihoods and skills; but they point to some serious misconceptions about the nature of technology diffusion and social organization in farming communities and suggest significant failings in the way that support for agricultural technology generation is conceived. The conclusions are presented in the context of a pragmatic approach to agricultural development and it is hoped that the analysis presented here can contribute to more effective and practical strategies in support of smallholder farming.

The research described in this book was supported by an Economic and Social Research (ESCOR) grant from the UK Department for International Development (DFID). Their support is gratefully acknowledged, and the usual disclaimers apply. Each of the three cases studies benefited from reviews by specialists who are acknowledged in the individual chapters. In addition, the initial conclusions from the research were presented and discussed at a small workshop called 'Human and Social Capital in Low External-Input Agriculture in London in June 2004. The participants in that workshop brought an exceptionally wide range of experience to the discussion and their suggestions and analysis were invaluable in refining the arguments that appear in this book. The workshop participants included Barbara Adolph, Alastair Bradstock, Elizabeth Cromwell, Chris Garforth, Jon Hellin, Catharine Moss, Christine Okali, John Pender, Colin Poulton, Barry Pound, Ian Scoones, Andrew Shepherd, Alistair Sutherland, Jim Sumberg, Dan Taylor, Janny Vos and Stephanie Williamson. While many of them would not necessarily subscribe to all of the book's conclusions, their contributions to clarifying issues and challenging particular interpretations are gratefully acknowledged.

Margaret Cornell provided excellent editorial support in the preparation of the book. Colleagues at ODI contributed to the management of the project and the production of the book's manuscript, and Leah Goldberg and Oliver Reichardt deserve special thanks.

List of Acronyms and Abbreviations

ACORDE	Co-ordinating Agency for Development (Honduras)
AEARP	Agricultural Extension and Adaptive Research Project (Sri Lanka)
AI	agricultural instructor
C	Celsius
CF	conservation farming
CGIAR	Consultative Group on International Agricultural Research
CIDICCO	International Centre for Cover Crop Information (Centro Internacional de Información Sobre Cultivos de Cobertura)
cm	centimetre
COSECHA	Association of Sustainable Agriculture Advisers (Honduras)
DFID	Department for International Development (UK)
DoA	Department of Agriculture (Sri Lanka)
DPT	divisional planning team (Kenya)
FAO	Food and Agriculture Organization of the United Nations
FFS	farmer field school(s)
FSR	farming systems research
GIS	geographical information system
GM	genetically modified
GRM	green manure
ha	hectare
ICIPE	International Centre of Insect Physiology and Ecology (Kenya)
ICRAF	World Agroforestry Centre
ICRISAT	International Crops Research Institute for the Semi-Arid Tropics
ILRI	International Livestock Research Institute
IPM	integrated pest management
IPPM	integrated production and pest management
IRT	in-row tillage (Honduras)
ITDG	Intermediate Technology Development Group
ITK	indigenous technical knowledge
KARI	Kenya Agricultural Research Institute
kg	kilogram
KICOBADE	Kiye Community-Based Development Project (Kenya)
lb	pound
LDP	Livestock Development Project (Kenya)
LEIA	low external-input agriculture

LEISA	low external-input sustainable agriculture
LEIT	low external-input technology
LISA	low-input sustainable agriculture
m	metre
ME	marginal effect
mm	millimetre
MOP	muriate of potash
NALEP	National Agriculture and Livestock Extension Programme (Kenya)
NDDP	National Dairy Development Project (Kenya)
NGO	non-governmental organization
NSWCP	National Soil and Water Conservation Programme
ODI	Overseas Development Institute
OR	odds ratio
PLAR	participatory learning and action research
PRA	participatory rural appraisal
Sida	Swedish International Development Agency
SL	sustainable livelihoods
SRI	system of rice intensification
SWC	soil and water conservation
TSP	triple superphosphate
T&V	Training and Visit system
UK	United Kingdom
US	United States
WN	World Neighbors

Low External-Input Technology (LEIT) and Agricultural Development

Robert Tripp

External inputs and agricultural development

The 1960s was a time of great hope for agriculture in developing countries. It marked the beginning of what became known as the Green Revolution, the principal manifestation of which was the spread of short-strawed, fertilizer-responsive varieties of wheat and rice that led to vast increases in food supplies in many Asian countries. For a while it looked as if the strategy of supplying appropriate varieties and complementary fertilizers, pesticides and other inputs could bring an end to rural poverty and chronic food shortages. The formula was based on provision systems, often subsidized, for external inputs, and relied upon technology packages whose practices were similar to those used in industrialized countries.

At the same time, however, another 'revolution' was also beginning to take shape, marked by the publication in 1962 of Rachel Carson's *Silent Spring*, which provided an inventory of the environmental and health consequences of an ever-growing dependence upon insecticides in agriculture and elsewhere. The book helped to galvanize an environmental movement that focused first on the effects of inputs in industrialized agriculture and then increasingly turned its attention to the situation in developing countries.

There are few examples of early confrontations between the two campaigns since each gained strength in relatively restricted spheres. The environmental movement dedicated much of its early concern to the consequences of pollution and dependence upon chemicals in industrialized countries. Meanwhile, the Green Revolution met its first critics not from the environmental side, but rather among those who detected a bias in its beneficiaries. There were concerns that the new technology was mostly of use to larger farmers, who were thought to gain at the expense of their poorer neighbours (e.g. Griffin, 1974). Although much evidence of fairly rapid catch-up by smaller farmers in the same communities became available, the concern that it was mostly farmers in favoured environments who benefited from such technology was more difficult to refute (Lipton with Longhurst, 1989).

As agricultural development programmes tried to redress this imbalance and to devote resources to technologies appropriate for more challenging environments, the top-down nature of the Green Revolution approach was increasingly called into question. It was recognized that farmers in 'complex, diverse and risk-prone' areas (Chambers, 1997) could not take advantage of standardized packages of practices and that other strategies for technology generation would be required. Considerable resources went towards the development of more adaptive and location-specific research techniques, exemplified by the farming systems research (FSR) movement (Collinson, 2000). By the 1980s, adaptive research programmes were paying increasing attention to the role of farmers in technology generation, and the focus shifted towards strengthening farmers' skills and responsibilities in the process, leading to a significant expansion of interest in various types of farmer participatory research (Chambers et al, 1989). The emphasis on farmer participation (combined with declining support for public research and extension) spurred a marked increase in the involvement of non-governmental organizations (NGOs) in agricultural technology programmes. The NGO activity was sometimes seen as simply a more efficient alternative to badly organized state agricultural agencies; but it often involved more fundamental concerns of many NGOs for local empowerment and for bringing environmental concerns to the fore in agricultural development (Farrington and Bebbington, 1993).

By this time, environmental problems caused by dependence upon external agricultural inputs were well documented for both industrialized and developing countries (Conway and Pretty, 1991). In the South, pesticide use was increasing in response to both Green Revolution programmes and the marketing campaigns of the industry. Pesticide misuse was increasingly common and public health concerns were apparent. Fertilizers were an increasingly important component of agricultural development programmes, and although they offered higher yields, they sometimes caused farmers to abandon traditional techniques of soil fertility maintenance. Access to irrigation grew, bringing with it new possibilities of crop production, but also instances of salinization and threatened water tables. Agricultural expansion and mechanization were, at times, the cause of increased soil erosion, and constraints such as the appearance of plough pans became more common. Not only did increased use of inputs raise questions about environmental sustainability, but the costs of the chemicals, irrigation and mechanization were often subsidized, raising further concerns about the capacity to support these strategies in the long term.

The combination of growing environmental awareness, the increasing role of NGOs in donor-led agricultural development and the realization that, in many cases, agricultural progress depended crucially upon local knowledge and local solutions was responsible for a significant shift in attitude regarding agricultural technology. External inputs were again the focus; but this time the goal was to reduce their role in the development of improved farming methods. Such a shift would not only mitigate further environmental degradation, but would also help to build up local capacities and promote the development of well-adapted farming methods. The shift was not marked by the emergence of a single, monolithic movement, but rather represented a significant confluence of the interests and priorities of various groups.

This brief summary cannot substitute for a more detailed and competent history of agricultural development efforts over the past 40 years; but it is hoped that it provides a sufficient outline of some of the major factors that challenged the Green Revolution's strategy, based on external inputs. The performance of and prospects for alternatives to the use of external inputs in agricultural development are the focus of the concerns of this book.

Sustainable agriculture

The term most commonly used in framing a response to the environmental and equity challenges posed by dependence upon external inputs is 'sustainable agriculture'. Unfortunately, the term seems to create as many problems as it resolves. It has come to be used in such a wide range of contexts and with such variable meaning that it often causes confusion. The term includes both an approach to evaluating agricultural development and a set of specific techniques and technologies. The approach features a number of concerns (highlighted in varying degrees by those who use the term), including an emphasis on the externalities of agricultural production (particularly environmental costs), a belief that the use of conventional discount rates for assessing the returns to investment in agriculture represents a 'dictatorship of the present', and a focus on the regeneration of the countryside and rural social systems. With respect to technological approaches, the range of those espousing support for sustainable agriculture includes (at one extreme) those seeking to eliminate the use of all external ('synthetic') inputs from farming and (at the other extreme) those who are simply willing to acknowledge the need for addressing certain environmental excesses in the pursuit of otherwise conventional and 'high-tech' agricultural development strategies. As Conway (1997, p163) observes: 'Almost anything that is perceived as "good" from the writer's perspective can fall under the umbrella of sustainable agriculture.' The identification and specification of measures of sustainability have occupied a number of authors, and there is no indication that any broadly accepted agreement on analytical tools will soon emerge (Douglass, 1984; Tisdell, 1988; Yunlong and Smit, 1994). A precise and unambiguous definition of sustainability is probably unreachable (e.g. Pretty, 1995) and we shall not attempt any further elaboration in this book.

Despite this lack of clarity, sustainable agriculture is a widely recognized description of a set of strategies and aspirations that are sufficiently coherent and widespread to warrant careful evaluation. An acceptably robust approximation of the major elements is provided by Pretty (2002, p56), who presents sustainable agriculture as:

> ... farming that makes the best use of nature's goods and services
> while not damaging the environment. Sustainable farming does
> this by integrating natural processes, such as nutrient cycling,

nitrogen fixation, soil regeneration and natural pest control, within food production processes. It also minimizes the use of non-renewable inputs that damage the environment or harm the health of farmers and consumers. It makes better use of farmers' knowledge and skills, thereby improving their self-reliance, and it makes productive use of people's capacities to work together in order to solve common management problems.

Although there will always be debates over the details, these elements are widely supported by a growing list of authors, practitioners and observers who may describe their interests as sustainable agriculture, low-input agriculture or agro-ecology. An important characteristic that distinguishes these movements and programmes is an interest in lowering the reliance on external inputs, not only to reduce environmental damage, but also to take better advantage of local biological, human and social resources. The shift away from external inputs is often seen as not merely a choice of technology based on ecological considerations, but a strategy that has profound consequences for the economy, social organization and purpose of farming communities. No one claims that the pursuit of such far-reaching aspirations depends simply upon shifting farmers' patterns of input use; but technological alternatives to external inputs have assumed a role well beyond environmental protection and are presented as a crucial element in the development of more self-reliant and equitable rural communities.

The breadth of interest in sustainable agriculture

The interest in reconsidering agriculture's dependence upon external inputs is certainly not confined to the South. A significant proportion of current debates regarding European farming are framed in the context of sustainable agriculture, and demand for organic produce is booming in the supermarkets. The shift of focus is also evident in the US, although with a somewhat lower profile. A comprehensive study describes the types, growth, and prospects of 'alternative agriculture' in the US (NRC, 1989). The practice of alternative agriculture is distinguished by a number of characteristics, including 'more thorough incorporation of natural processes such as nutrient cycles, nitrogen fixation and pest–predator relationships into the agricultural production process; reduction in the use of off-farm inputs with the greatest potential to harm the environment or the health of farmers or consumers ... [and] improvement of the match between cropping patterns and the productive potential and physical limitations of agricultural lands to ensure long-term sustainability of current production levels' (NRC, 1989). The report concentrates on the technological and economic characteristics of alternative agriculture; but much of the interest in alternative agriculture in the US is linked to hopes for a renaissance of the small family farm (Buttel et al, 1986).

The significant shift in the perception of agricultural progress is illustrated by a story told by Everett Rogers (1995), who has dedicated a career to synthesizing research on the spread of innovations in a variety of fields, including agriculture, and trying to explain the differences between 'innovators' and 'laggards'. His own thesis research, conducted in Iowa during the early 1950s, examined the uptake of inputs such as herbicides, fertilizers and veterinary antibiotics. One farmer in his sample rejected all such technologies because of environmental concerns. 'My research procedures classified him as a laggard in 1954; by present day standards he was a super-innovator of organic farming' (Rogers, 1995, p186).

In developing country agriculture, an undeniable shift in priorities and perceptions regarding agricultural sustainability has led many observers to envision nothing less than another revolution in the making. Greenpeace urges the development of a:

> ... coherent international movement which is capable of provid-
> ing an alternative to the conventional system. As ecological
> agriculture becomes more successful economically, and an
> increasing number of farmers throughout the South decide –
> independently or with assistance from NGOs [non-governmental
> organizations] – to jump off the chemicals treadmill, the chances
> of this real Green Revolution succeeding become greater every
> day (Parrott and Marsden, 2002).

Madeley (2002, p41), summarizing a number of reports on sustainable agriculture and organic farming, concludes that a 'revolution is taking shape in developing country agriculture'.

The degree to which such a revolution is actually in progress is one of the principal concerns of this book. A review of much recent NGO literature would indicate that the initial barriers have been breached. It is increasingly common to read glowing reports of the impact of the new strategy. For instance, sustainable agriculture programmes in one area of Nepal are credited with having 'transformed areas that used to suffer food shortages into ones that now produce an abundance of honey, fruit, cereals, rice and leafy greens ... and the communities have diversified into cottage industries, including bee keeping, cotton and hemp handlooms, leather processing, candle-making, agroforestry and kitchen gardening' (Orton, 2003, p42). Almost everyone would welcome this type of transformation but we need to know the degree to which technological change can spearhead such a remarkable metamorphosis in some of the world's poorest farming communities.

Needless to say, this optimistic vision is not universally accepted, and there are concerns about a reliance upon low-input strategies from both economic and biological standpoints. Low (1993, p98), commenting on the promotion of low-input sustainable agriculture (LISA) in Southern Africa, warns that such 'prescriptions, which imply increased inputs of scarce household labour time, are shown not to be consistent with past adoption behaviour of farm households. Unless the socio-economic circumstances in which farm households operate change, or are changed, it is unrealistic to expect that households will swallow a LISA-type pill, even though it may improve environmental health'. The father of the Green Revolution, Norman Borlaug (cited in Haggblade et al, 2004, p3), also commenting on African agriculture, dismisses the biological assumptions of low-input strategies and contends that 'the most environmentally friendly action that can be taken in sub-Saharan Africa is to promote moderate and proper use of chemical fertilizers in an aggressive manner... There simply is not enough organic fertilizer available to provide sufficient nutrients to the soil to satisfy the growing food demand of Africa.'

Sustainable agriculture and traditional farming

Agriculture is unique among humankind's enterprises with respect to the emotions that it evokes. Agricultural practices are inextricably woven into broader traditions, customs and wisdom. A particularly effective argument for more attention to agricultural sustainability is based on comparisons with traditional farming practices. On the one hand, many agricultural systems have taken millennia to evolve, and their intensification because of population growth and the availability of external inputs may upset environmental balances that will be difficult to restore. Land shortages have destroyed many traditional fallowing practices and have cut farmers off from other common property resources. Economic pressures and opportunities have made farmers look for additional ways to reduce their labour commitments on the farm. External inputs are often cheap and convenient, and their distribution may benefit from various direct and indirect subsidies. The consequences of these immense changes in resource use in agriculture are difficult to predict:

> [The] historic brevity of modern intensive agriculture should make us cautious when assessing its long-term capacities. Such regions as China's Huang Valley or Iraq's Tigris-Euphrates alluvium have been farmed continuously by traditional methods for more than 7000 years (Smil, 2000, p65).

On the other hand, an idealization of traditional methods overlooks the fact that some of the environmental problems we observe today are 'merely continuations and intensifications of undesirable changes that had nearly always accompanied traditional cropping' (Smil, 2000). A review of China's premodern agriculture warns of the dangers of romanticizing a system whose exploitation of forests and upland soils made it ultimately unsustainable (Elvin, 1993). Ellis and Wang (1997) describe an exceptionally complex and productive farming system in an area of China where, by the mid-19th century, poverty was so severe that peasants pawned their winter clothes in summer in order to buy grain, repurchasing them after the autumn rice harvest. The contradictions in the system are illustrated by the fact that Chinese agriculture today is highly dependent upon external inputs:

> ... survival of peasants in the ricefields of Hunan and Guangdong – with their timeless clod-breaking hoes, docile buffaloes and rice-cutting sickles – is now more dependent on fossil fuels and modern chemical syntheses than the physical well-being of American city

dwellers sustained by Iowa or Nebraska farmers cultivating sprawling grainfields with giant tractors (Smil, 1991, p593).

Neither side in the argument pretends that the issue is merely what type of technology is best for farmers, and distinct visions of the nature of farming communities are also at stake. Until recently, of course, all farmers relied upon local resources, and the promotion of a more self-sufficient agriculture in the 21st century often draws upon images from an earlier era. As an illustration of the changes occasioned by the Green Revolution in Asia, Pretty (2002, p36) describes the impact on the Balinese irrigated rice system:

> Soil fertility was maintained by the use of ash, organic matter and manures; rotations and staggered planting of crops controlled pests and diseases; and bamboo poles, wind-driven noise-makers, flags and streamers scared birds. Rice was harvested in groups, stored in barns and traded only as needs arose. The system was sustainable for more than 1000 years. Yet, in the blink of an eye, rice modernization during the 1960s and 1970s shattered these social and ecological relationships by substituting pesticides for predators, fertilizers for cattle and traditional land management, tractors for local labour groups, and government decisions for local ones.

The Indonesian agricultural policies that supported the aggressive promotion of seed-fertilizer technology had undeniably profound impacts upon farming communities such as these. But whether a centuries-old ecological and social balance was overturned 'in the blink of an eye', or whether pre-Green Revolution farming communities were generally as idyllic as the one described, can be debated. Geertz's (1971) study of ecological change in Indonesia provides descriptions of most of Java and southern Bali in the late 19th and early 20th centuries under Dutch colonial rule, where by 1900 rice production could not keep pace with population growth and the demands of the colonial sugar industry. In most areas, per capita production dropped a further 20 per cent between 1920 and 1955, and Indonesia was fast becoming one of the largest rice importers in the world. In the villages, the relatively equitable distribution of ever-smaller plots of land was mediated by complex systems of sharecropping and exceptionally high investments of labour per hectare, the 'involution' that provides Geertz with the theme of his study. Such involution led to a 'shared poverty' in the villages, where life was 'complicated in the diversity, variability, fragility, fluidity, shallowness and unreliability of interpersonal ties, [and] simple in the meagre institutional resources by which such ties were organized' (Geertz, 1971, p103). Rather than a land of self-sufficient farmers, the colonial economy led to a situation where 'the Javanese cane worker remained a peasant at the same time that he became a coolie, persisted as a community-oriented household farmer at the same time that he became an industrial wage labourer. He had one foot in the rice terrace and the other in the mill' (Geertz, 1971, p89).

Many such examples of contrasting images of peasant life could be extracted from the literature. They are not necessarily contradictory and their juxtaposition merely underlines what the parties in the debate choose to emphasize. Just because peasants were perhaps less self-sufficient and productive in the past than we might like to think does not mean that they do not deserve a better future; and just because their livelihoods were not exclusively dedicated to subsistence agricultural production does not mean that they did not exercise exceptional creativity in making use of local farming resources. But similarly, because farmers utilized fewer external resources did not necessarily make them self-sufficient; and the further back into colonial and pre-colonial societies that we must search for truly independent subsistence production, the less relevant these examples are as models for the future.

Variations on the theme

There are a number of approaches to the theme of sustainable agriculture, and we shall need to select a particular focus for the analysis that follows.

One prominent interpretation is described as agro-ecology. Those working on agro-ecology include both academic researchers and development project practitioners (Altieri, 1995; Uphoff, 2002a) and are concerned about production systems rather than particular technologies. Agro-ecology recognizes that an agricultural system differs from a natural ecosystem because it includes external controls and sources of energy and lowered diversity; but it approaches the study of field crop cultivation as a system involving ecological processes such as nutrient cycling and predator–prey interactions. It also includes a focus on how social factors contribute to system stability and resiliency (Altieri, 1995). The issues of stability and resilience are of particular importance, and one of agro-ecology's principal concerns is the vulnerability of oversimplified modern agricultural systems that rely upon artificial, external controls.

Another important example is represented by organic agriculture. In most industrialized countries (and in a growing number of developing countries), organic produce is defined according to a strict set of criteria regarding production conditions, including the elimination (with few exceptions) of all external inputs, particularly mineral fertilizers and synthetic pesticides. Formal systems of organic production are aimed at markets where produce is subject to a strict certification system and, in turn, is likely to receive a price premium from consumers who are concerned about the health risks of conventionally produced crops and/or who wish to support more environmentally sustainable production.

There are interpretations of sustainable agriculture which are even more comprehensive than that for organic agriculture, such as permaculture, which envisions farm production in a holistic and integrated fashion, aims for self-sufficiency, includes attention to compatible technologies for housing and energy provision, and offers its own strategies for community development. Although such comprehensive approaches are most often attempted in industrialized countries, fairly strict philosophies discouraging external input use can be found in various projects in the South. Rigid adherence to such ideologies may, at times, hinder progress in developing relevant technologies for resource-poor farmers, as Vaessen and de Groot (2004) report for a project working on soil fertility enhancement and soil and water conservation in highland Guatemala.

Some interpretations of sustainable agriculture argue that positivist science may not be capable of developing acceptable solutions for the problems faced by agriculture around the world. Altieri (1995) believes that the rise of positivist science is partly responsible for the loss of respect for agro-ecological systems and the lack of interest in holistic approaches. Pretty (1995) argues that it is incorrect to see science as a value-free and abstract search for truth, and believes that its reductionist approach is not capable of accommodating the variability and context specificity demanded for pursuing more sustainable production systems. This is not to say that an interest in sustainable agriculture is anti-science (and most critics of positivist science nevertheless fill their books with reports from conventional scientific studies), but rather that many versions of sustainable agriculture see the scientific method as only one tool for developing more equitable and sustainable agricultural technology.

Despite an association with what may appear to be some fairly radical or rigid conceptions of agricultural development, however, the vast majority of approaches to sustainable agriculture in fact share a much more pragmatic view of technology, and relatively few espouse a total rejection of external inputs. Instead, there is recognition from most quarters that there are many important reasons for placing more emphasis on the use of local resources and skills as part of agricultural technology development.

The stance adopted here supports the view that the search for strategies to improve small farm productivity and stability must be conducted as widely as possible.

The degree of support for a pragmatic view is evident in statements from a wide range of observers:

> ... it is time to end the artificial conflict between so-called 'modern' methods, based on chemical fertilizers, irrigation and improved cultivars, and 'traditional' or 'agro-ecological' methods, based on intercropping, rotations, cover crops and organic nutrient supplements... Productivity improvements depend on recognizing and reinforcing complementarities (Barrett et al, 2002).
>
> Real progress can only come from a synthesis of the best of the past, eliminating practices that cause damage to environments and human health, and using the best of knowledge and technologies available to us today (Pretty, 2002, pxiii).

This book focuses on experiences with efforts in technology development aimed at promoting low external-input alternatives in the context of practical programmes for improving the welfare of smallholder farmers.

Low external-input technology (LEIT)

The examination of low external-input agriculture must acknowledge that the issues at stake go well beyond the use of particular production resources. The concerns include the capacities of farmers and their communities and the nature of agricultural development. Arguments are often based on interpretations or aspirations that reflect a range of values and beliefs. This book does not attempt to examine those underlying beliefs; but while focusing on a particular type of technology (involving a minimal use of external inputs), it also pays attention to the concomitant issues of farmer capacities, resources and organization.

If our starting point is a particular set of agricultural technology options, it will be useful to refer to these with an appropriate term. We need to define the technologies that we shall later examine, both for their contributions to improving farm productivity and their roles in strengthening farmer and community capacities.

We are interested in the growing range of agricultural technologies that feature the use of local inputs and resources, consider long-run environmental consequences as well as short-run production gains, and eschew top-down recommendations in favour of bottom-up adaptation. The term 'sustainable agriculture' is so broad and embodies so many (at times contradictory) interpretations that a substitute is required. Because our focus is first on particular types of technology and then on their wider consequences related to farmer and community capacities, we shall use the term 'low external input'. This choice is not without problems. Agriculture that relies upon low external inputs does not necessarily address environmental concerns, promote farm-level adaptation or promote community organization. These considerations have been partly addressed in the literature, and the earlier term 'low external-input agriculture' (LEIA) has been largely replaced by 'low external-input sustainable agriculture' (LEISA) (Reijntjes et al, 1992), the latter being defined as 'agriculture which makes optimal use of locally available natural and human resources … and which is economically feasible, ecologically sound, culturally adapted and socially just' (Reijntjes et al, 1992, pxviii).

While not disputing the merits of the goals included in this definition of LEISA, this book cannot attempt an evaluation of such a comprehensive vision of agricultural change. Instead, it takes as its focus the development of a range of technologies that complement or substitute for external inputs (and, hence, may be more accessible to farmers), provide significant environmental benefits, and usually require local adaptation. The goal of the book is to look at the farm-level utilization of this type of technology and to understand the correlates of farm management, farmer capacity and community organization that relate to its performance; because established terms often bring with them potential associations and stances, it will be better to adopt as neutral a description as possible for the type of technology under examination. Throughout the rest of the book descriptions of these types of technologies will use the term low external-input technology (LEIT).

Defining LEIT

It is difficult to draw precise boundaries around the category of practices that we shall call LEIT. One of the issues is the crop management context in which a particular practice is carried out. Many proponents of LEIT take a pragmatic approach and consider instances where LEIT may complement external input use. However, some of the more enthusiastic proponents of sustainable agriculture often ignore the wider crop management context in which a particular example of LEIT is found, mistaking important instances of innovation in local input use or environmental conservation for complete conversion. For instance, Asian rice farmers who use integrated pest management (IPM) to reduce insecticide use in rice or Brazilian farmers who practise conservation tillage are often counted as examples of 'sustainable agriculture' despite the high levels of fertilizers and herbicides they employ on their farms. Similarly, a recent example of intercropping two different rice varieties in China to control rice blast disease (in place of fungicide application) reported in the journal *Nature* (Zhu et al, 2000) has been used by several NGOs to highlight the wisdom of traditional practices, ignoring the fact that the experiment took place in a production system relying upon hybrid seed and some of the highest external input levels in the world.

Another issue in the definition of LEIT involves an understanding of what constitutes 'external' inputs. LEIT does not always imply a strict reliance upon farm-level resources, and it is best to remember that the major motivation is to reduce reliance upon non-renewable resources rather than to limit trade in agricultural inputs. Among the most common targets for reduction or replacement are manufactured pesticides (including insecticides, fungicides and herbicides) through various examples of IPM. But IPM may include a number of 'natural' technologies (for example, bio-pesticides and pheromone traps) that must usually be purchased in the market. Another target for LEIT is mineral fertilizer; although most organic soil amendments are produced on the farm, there is also a significant local market for some of these inputs, where the landless or poorer farmers may collect manure or produce composts upon behalf of wealthier farmers (Butterworth et al, 2003). Conversely, there are important examples that promote the elimination of tillage operations even though these may be carried out with farmers' own traction resources, the rationale here being resource conservation and the promotion of complementary LEIT practices.

A further concern for the definition of LEIT for the analysis that follows is the origin of the technology. Until recently, of course, everyone was a LEIT farmer. Discoveries at the end of the 19th century enabled the chemical industry to produce mineral fertilizers; but they were not in widespread use until the middle of the 20th century. Similarly, although synthetic pesticides have been manufactured and used for more than a century, their application accelerated only after World War II. Although most farm-level innovation in the South continues to include examples of LEIT, we are not going to look specifically at local traditions and will concentrate, instead, on experiences of introducing examples of LEIT that are novel to particular farming systems or at least encourage significant elaboration on local practices.

LEIT is, of course, not confined to small-scale agriculture in developing countries. Technologies such as IPM and conservation tillage are also part of many large-scale commercial farming operations. Moves towards precision farming (Cassman, 1999) often aim to reduce external input use and require detailed information and management. Uphoff (2002c) notes the similarities in principles and outcomes between a novel, low-input system of rice management for resource-poor farmers in Madagascar and newly developed methods of raised-bed cultivation for wheat in a much more technologically advanced area of northern Mexico. However, the examples in this book will be largely confined to small-scale agriculture.

Finally, we need to acknowledge that seed plays an anomalous role in LEIT. On the one hand, it may be seen as a purchased input, although ultimately based on farm-level resources (germplasm). Modern crop varieties are, of course, a major component of high-input agriculture; many modern crop varieties are more fertilizer responsive than local varieties (although this does not imply that they are necessarily fertilizer dependent) (Tripp, 1996). Some proponents of sustainable agriculture favour the use of local crop varieties (or at least reliance upon local seed systems); but there is a wide range of opinion and many who promote LEIT welcome the introduction of new crop varieties as long as they are appropriate to local farming systems (e.g. Reijntjes et al, 1992). There is a stronger correlation between support for LEIT and rejection of genetically modified (GM) crop varieties, although more nuanced approaches to the subject can also be found (e.g. Pretty, 2003). This book does not consider any particular seed or variety strategy as an integral part of LEIT, and makes little reference to seed practices or variety use in the following chapters.

In summary, the book is not based on an airtight definition of LEIT. While recognizing that the subject of low-input agriculture attracts devotees, purists, sceptics and detractors, the book does not aim to defend or dissect any particular set of views. LEIT may be seen in a negative (the reduction or elimination of external inputs) or positive (the promotion of local resources and skills) sense, and the reader is free to choose. The principal motivation for initiating this study is that the types of technology under examination have considerable potential. The poverty levels of many farm households preclude any reasonable hope that they could take advantage of technologies requiring significant financial investment; the logic of making better use of locally available resources is unquestionable. Environmental problems of resource degradation and pollution from chemical inputs are serious challenges that must be addressed. The realization that most technology needs to be adapted to individual farm circumstances is an additional justification for promoting innovations based on local resources and skills, and the development of such resources and skills certainly offers opportunities for empowering farmers and their communities.

It is hoped that the analysis in this book will be relevant to a wide range of people concerned with pro-poor agricultural development, ranging from those who see LEIT as simply one part of a larger strategy for improving farm productivity, to those who are convinced that LEIT is the key to a fundamental shift in the nature of small-farm agriculture and the development of farming communities. The book's primary focus is a pragmatic one, looking first at individual technologies and their performance, followed by an examination of wider impacts upon farmer capacities and organizations. It does not pretend to analyse the broader principles or values regarding the nature of development or smallholder farming that are often part of discussions on sustainable agriculture. The description of this approach as pragmatic is not meant to imply any type of special objectivity. All analysts have their biases and these affect the way in which they conduct their work and draw their conclusions. The approach adopted in this book is that a certain class of technology (LEIT) has been promoted with a set of expectations regarding its immediate and secondary effects on farming communities. Those expectations represent worthwhile objectives, shared by people espousing a range of beliefs, and it is important that we assess current performance as carefully as possible and draw conclusions for further actions.

Issues related to the performance of LEIT

Several questions need to be addressed if we are to understand the performance and potential of LEIT. One of these involves assessing the economic and environmental benefits of LEIT. This is obviously a key question; but its answer lies largely outside the scope of this book. A second question involves the actual extent of uptake and spread of LEIT; there are widely differing opinions on this subject and the book hopes to provide relevant information on this theme. A third set of questions concerns the costs of LEIT; these technologies are often considered to be labour- and information-intensive and the book will explore the relevance and accuracy of such characteristics. A fourth set of questions that the book will address is related to the expectation that the spread of LEIT can serve to strengthen human and social capital in farming communities.

Because LEIT is seen as not only offering acceptable economic returns, but also providing longer-term environmental benefits, its evaluation is particularly problematic. Measuring and valuing the environmental improvements (for farmers and society), and deciding on an appropriate discount rate to apply to the assessment of benefits that may accrue over a period of many years, are difficult challenges that feature in much of the literature on LEIT. A recent attempt to assess the yield gains from the uptake of 'sustainable agricultural practices and technologies' (including, but not limited to, what we define as LEIT) examined data from 89 projects and found an average per project increase of 93 per cent in food production (Pretty et al, 2003). Although this is the most comprehensive attempt to date, the conclusions are based almost entirely on the reports of the projects themselves and offer no independent means of verification. Graves et al (2004, p475) point out that 'despite the continuing debate on the relative performance of the two approaches, there are few studies that compare yields and production under the same soil and climatic conditions and over wide areas'.

The uncertainties about the outcomes and performance of LEIT are not confined to the South. Buttel et al (1986, p354) comment on two sets of opposing literature in the US regarding the benefits of reduced-input agriculture and ask: 'How can presumably capable, well-trained scientists who publish papers in respectable, refereed journals and other publications come up with such diametrically opposed results?'. They conclude that part of the answer lies in the fact that proponents base their conclusions on small-scale studies without considering the feasibility of scaling up, while critics tend to base their arguments on simulations and models without careful field-level analysis of alternative practices.

The debate regarding the nature and extent of benefits from particular examples of LEIT will certainly continue for many years. In the meantime, one proxy for better statistics and more sophisticated analysis is the judgement of farmers themselves, reflected in the extent to which they utilize LEIT. The rates of adoption, adaptation and diffusion of LEIT provide some indication of the benefits (at least in the short term) of these technologies. In addition, LEIT's aspirations to reach farmers and environments overlooked by conventional technology mean that an examination of the type of farmers who take advantage of LEIT is as important as merely documenting the frequency of use.

This book will review what is known about the utilization of LEIT and will offer original case studies that further our understanding on the subject. There is already some literature on the uptake of LEIT; but it exhibits a number of weaknesses. First, there is simply not enough of it. Because LEIT is a rather new endeavour and its applications are generally limited to location-specific projects, there has not been the degree of follow-up that characterizes the evaluation of much conventional agricultural technology (there are a few notable exceptions, such as the extensive work of the World Agroforestry Centre, or ICRAF, in analysing the uptake of innovations in agroforestry – e.g. Franzel, 1999; Place et al, 2002a). Second, much of the literature on the uptake of LEIT is based on self-reports and impressionistic assessments. The most comprehensive review (Pretty et al, 2003) estimates that sustainable agricultural practices can be found on at least 3 per cent of the arable land in Africa, Asia and Latin America (but, given that the majority of the area covered in these examples is on large farms, the proportion of farmers utilizing these techniques would be significantly less). These adoption estimates are provided largely by the projects themselves. In most other cases as well, we simply have no independent assessment of the extent of uptake of particular examples of LEIT. A third reason for the paucity of hard data is the fact that much LEIT is promoted either through short-term projects where little provision is made for comprehensive evaluation or by NGOs with limited funds and incentives for follow-up (Altieri, 1995). In addition, even when formal studies regarding LEIT adoption have been conducted, they tend to focus on the immediate aftermath of intensive investments in location-specific projects that have benefited from high-quality extension (Place et al, 2002b). By taking a more long-term and independent view of the uptake of LEIT, this book hopes to remedy some of these deficiencies.

Low-input strategies are often seen as being particularly appropriate for the rural poor, both those who farm in marginal environments and those who are at the bottom of the village's economic ladder. In discussing appropriate policies for less-favoured farming areas, Kuyvenhoven (2004, p417) sees possibilities for rapid improvements in crop yields based on simple technology transfer and farmer innovation, focusing on low external-input farming 'using crop rotations, improved fallows and manuring practices'. Not only does such technology offer promise for difficult environments, but many see it as helping to correct the imbalance of most conventional technology:

> ... these approaches could enhance human resources by improving people's management capacities at the same time as redressing disparities in distribution because agro-ecological methods are well suited for use by less-favoured households (Uphoff, 2002d, pxvi).

Altieri (1995, p175) believes that 'the poorer the farmer is, the more relevant low-input approaches are, given that poor farmers have little choice but to use their own resources'. Any analysis of LEIT needs to pay particularly close attention to how patterns of uptake are related to the distribution of poverty.

LEIT is often characterized by both its proponents and critics as being labour- and information-intensive. LEIT may use labour in place of purchased inputs to address crop management challenges; the labour may be provided by individual farmers or group action. In many cases the technology involves new information that farmers can use to strengthen their knowledge base. Observers' contrasting interpretations of LEIT depend largely upon whether they believe that labour and information are readily available to resource-poor farmers, and whether a particular type of technology can develop and strengthen local labour resources and build up new knowledge. But labour and knowledge are important for many types of technology, and there are legitimate questions as to whether LEIT is, in fact, a special case in this regard. The book will look at these issues in some detail.

Although labour and information are widely treated in the literature on agricultural technology, they assume particular importance for LEIT and are often included in broader aspirations related to strengthening human and social capital. There is a feeling that locally adapted technologies 'are most likely to emerge from new configurations of social capital, comprising relations of trust embodied in new social organizations, and new horizontal and vertical partnerships between institutions, and human capital comprising leadership, ingenuity, management skills and capacity to innovate' (Pretty et al, 2003, p219). Because LEIT is based on principles rather than recipes, it promotes an adaptive and innovative attitude. Some authors see these skills as having very broad application. Uphoff (2002b, p16) holds that 'farmers who have developed their analytical skills and their confidence from agricultural experimentation will be better able to be productive in cities if or when they are displaced to an urban environment'.

Because LEIT relies upon local resources, it requires particular knowledge, skills and organization. Despite the centrality of these characteristics to the promotion of LEIT, there has been very little formal examination of the human and social capital correlates of its utilization or analysis of the extent to which its successful use leads to improvements in farmer capacities for innovation, experimentation or information transmission, or improvements in community capacities for organization and further agricultural development. This book addresses these issues.

Questions about the performance of LEIT

The objectives of the book are to review the literature and contribute original case study material to address a series of questions about LEIT:

- First, we need a better understanding of the actual extent of its utilization. What proportion of farmers has been able to take advantage of particular examples of LEIT, and how extensively are the technologies used on the farms of the adopters?
- There is considerable hope that LEIT can reach many of the farmers and farming environments that have not profited from conventional agricultural technology programmes. Do the projects that are established to develop and diffuse LEIT elicit participation from a sufficiently wide range of farmers? Do the projects include the poorer members of the farming community and farmers in more marginal environments?
- Even when projects achieve broad participation, they need to offer techniques and methods that farmers can incorporate within their own crop management strategies. When examples of LEIT are taken up by farmers, are the adoption patterns significantly different from those of conventional technology? Does the nature of LEIT mean that the common biases in the adoption of conventional technology (towards better-resourced, more commercial farmers) are less in evidence?
- The labour component of LEIT is of particular importance. All things being equal, farmers prefer technologies that are labour saving. How accurate is it to characterize LEIT as necessarily labour intensive, and to the extent that examples of LEIT imply some additional labour investment, how is that accommodated by farmers? With the growing appreciation of rural livelihood diversity, how does the labour component of LEIT interact with household labour deployment strategies?
- LEIT often involves learning new skills. What resources do farmers have that allow them to acquire the necessary information to manage the technology? How does the resulting knowledge enhance general farming capacity? Is LEIT necessarily knowledge intensive?
- Although we are treating LEIT as simply a set of technologies, these technologies can also represent a distinctive approach to farming. Does the nature or rationale of LEIT encourage farmers to make broader changes in their farm management? Does the successful utilization of particular examples of LEIT lead to the emergence of an identifiable 'LEIT farmer'?
- LEIT usually involves local adaptation to accommodate individual farm circumstances. The process of adaptation itself develops skills that can be

applied further. Is there evidence that successful utilization of LEIT enhances farmers' innovative or experimental capacities?

• Both the high costs of intensive LEIT projects and the aspirations for community-based development argue for support to farmer-to-farmer transmission of information, either through the emergence of formally recognized farmer extensionists or through informal networks of information flow. When LEIT projects have successfully introduced new techniques to participating farmers, to what extent do those techniques, or the principles on which they are based, spread from farmer to farmer?

• When LEIT development and adaptation involve group activity, there are opportunities to build on this experience and take advantage of enhanced social capital. What is the experience of LEIT projects with respect to building up social capital?

The nature of the analysis

This book is focused on the performance of a particular class of agricultural technology – LEIT – in developing countries. It recognizes that the subject is the source of considerable controversy, but it is not going to examine claims regarding particular environmental or production benefits, nor is it going to endorse or challenge any particular vision of farming associated with discussions of LEIT. The book proceeds on the assumption that a significant proportion of the work on LEIT can offer real benefits to resource-poor farmers and that initiatives to take better advantage of local skills and resources and to strengthen farmer capacities and organization are well worth supporting. However, the attitude towards LEIT adopted here is empirical and pragmatic, rather than promotional or polemical. LEIT is not seen as a self-contained answer to all of the challenges of improving the livelihoods of resource-poor producers, but rather as an important contributor to a more comprehensive approach to rural development.

The analysis that follows is based on two types of information. The book reviews the literature, examining some of the relevant sources on the roles of labour and information in the development of agricultural technology, as well as related treatments on the roles of human and social capital, and looks in greater depth at the literature available on actual experience with various types of LEIT. The second source of information is a set of three original case studies from Honduras, Kenya and Sri Lanka that examine the experiences of several comprehensive and well-managed efforts at the introduction of LEIT.

The case studies

The three case studies were chosen to represent a wide range of examples of LEIT and to provide experiences from different farming environments. The cases share several characteristics. They all examine fairly extensive projects that covered many farming communities in order to avoid the biases inherent in assessing small pilot-project experiences. The cases were chosen from amongst well-managed LEIT projects, the objective of the analysis being to look at the outcomes of competently organized projects, not to examine problems in project management. In addition, the cases were chosen only after a desk and field review provided evidence that at least some of the examples of LEIT offered by the focus projects were appropriate to the environment and had been taken up by a significant number of farmers. Finally, the research was organized to examine the experience of participating farmers (and their neighbours) at least five years after the termination of project activity. Accurate assessment of technology adoption is often difficult in the immediate aftermath of a project, and delaying the assessment for a period of time allows for the examination of further technology adaptation, diffusion or abandonment, as well as providing opportunities to examine the evolution of project-related human and social capital.

The case studies examine the adoption of technologies introduced in the context of LEIT projects and follow their subsequent modification and diffusion. They pay particular attention to examining what type of farmer is best able to take advantage of these technologies and under what conditions. They also examine the extent to which access to labour or information determines technology utilization patterns. Finally, the timeframe of the studies allows for the examination of the consequences of the original project activities in terms of further innovation or organization.

This strategy of looking at the aftermath of fairly successful projects, five or more years after completion, has several strengths and weaknesses. One positive contribution is simply the fact that this type of study has not been done very often. Although project evaluations and assessments are frequently conducted, it is much less common to revisit a project well after completion. The strategy helps to reduce any direct influence of project resources and personnel and allows time for farmers to gain further experience with the principles and technologies that have been introduced. It also provides an opportunity to identify factors that are associated with technology utilization and diffusion, to assess the dynamics of farmer-to-farmer transmission of information, and to understand the development of human and social capital.

However, the strategy is not without its limitations. The identification of farmer assets at the time of the project and/or when the farmer first tested the technology is difficult, and fading memories mean that some of the original conditions and relationships that determined farmer interest or participation may be difficult to recall. The spread of information (and the movement of household members) means that it is difficult to establish a true control group for the studies that represents what farmers would have done had they not had access to project resources or information. Farming conditions and opportunities are continually changing, and the utilization of particular technologies must be viewed in relation to the evolving circumstances that affect the relevance of particular technologies, adding another dimension to the analysis. On balance, however, examining the long-term consequences of successful LEIT projects should make an important contribution to our understanding of the nature of technology adoption, adaptation and diffusion, as well as allowing a rare opportunity to assess the aspirations for LEIT's possible contribution to the enhancement of human and social capital.

The three case studies are based on quantitative surveys of technology utilization, supported by qualitative information obtained through informal interviews and observations. Some may question whether this type of formal methodology is appropriate for understanding LEIT, which relies heavily upon farmer adaptation and innovation rather than upon the acceptance of top-down prescriptions. These concerns are certainly valid, and studies of the adoption of agricultural technology are susceptible to mechanical management and interpretation. In addition, dividing the world into adopters and non-adopters can overlook the ongoing adaptation to farm- or field-specific circumstances that characterizes small-farm agriculture. The very nature of an adoption study can simply reinforce the top-down character of conventional agricultural development efforts by tallying the degree to which farmers follow the 'correct' recommendations.

However, these concerns are not sufficient to invalidate quantitative assessment of the outcomes of LEIT projects. In the first place, many of the internally generated assessments of such projects present quantitative results (number of farmers or size of area benefiting from a new technology, or number of groups formed). We simply need more independent sources of this type of information. In addition, the broader goals of LEIT projects, such as developing approaches to learning that improve problem-solving capacities or provide opportunities for collective action, must still be susceptible to measurement and evaluation if they hope to expect further support (Loevinsohn et al, 2002; Meinzen-Dick et al, 2004). Finally, we must recognize that embracing a wider set of criteria (participation and sustainable livelihoods) for the evaluation of rural development projects does not obviate the need for assessing specific instances of technical or institutional change.

Organization of the book

Chapter 2 outlines some of the major examples of LEIT. It does not provide a comprehensive review, but rather describes some of the most important examples of this type of technology, including methods for soil and water conservation, soil fertility enhancement, crop establishment and pest control. The chapter also summarizes the major strategies that are used to generate and promote LEIT, including group and participatory methodologies, farmer field schools (FFS) and other examples of social learning.

Chapter 3 reviews the role of labour and information in agricultural technology development. The discussion includes the organization of farm and off-farm labour (and the importance of livelihood diversity), human capital (including formal education and local skills), and the growing interest in the subject of social capital and its relation to agricultural change.

Chapter 4 reviews the literature on the performance of LEIT. It discusses what is known about the ability of LEIT to target poorer segments of the farming population, examines what is known about the importance of economic incentives in LEIT utilization, and reviews the roles of labour and human and social capital in the progress of LEIT.

Chapter 5 is a case study of a LEIT project in central Honduras. During the late 1980s and 1990s, several NGO projects promoted sustainable hillside farming through a range of methods for soil fertility enhancement and soil and water conservation. The study revisits two of the areas where the projects took place, examines the level of enduring adoption of techniques such as in-row tillage and the use of cover crops, assesses the reasons for variable adoption patterns, and looks for evidence of project contributions to human and social capital.

Chapter 6 is a case study of a soil and water conservation strategy in western Kenya. Kenya's Soil and Water Conservation Programme organized a nationwide Catchment Approach in which communities were encouraged to learn about and establish conservation techniques, including various types of terracing, vegetative strips and unploughed strips. The study examines the aftermath of the project in a series of villages in both high- and low-potential areas and places the results in the context of other soil and water conservation efforts in the area.

Chapter 7 is a case study of a farmer field school programme for integrated pest management in Sri Lanka. The Department of Agriculture, with assistance from the United Nations Food and Agriculture Organization (FAO), organized a nationwide FFS programme, starting in 1995. The study looks at the experiences of participants, neighbours and control communities in seven areas in southern Sri Lanka. It examines the extent to which participants follow different pest management and soil fertility practices, the nature of farmer-to-farmer information flow, and evidence of any type of further innovation or organizational development.

Chapter 8 summarizes the results of the three case studies and places these in the context of conclusions from the literature review. It reviews the uptake of the technologies, project participation, the patterns and correlates of adoption, the nature of information flow, the labour and knowledge requirements of the technologies, and evidence of further innovation or group activity. The chapter also discusses the implications of these results and suggests ways in which the promotion of LEIT could be made more effective. In particular, it questions the usual strategy of scaling up and suggests that more attention needs to be paid to scope, implying much better coordination of investments in LEIT in order to develop sustainable rural institutions.

Examples of LEIT and Farmer-Focused Development Strategies

Robert Tripp

Introduction

The purpose of this chapter is to review some of the major examples of low external-input technology (LEIT) and to outline the techniques most often used to promote this type of technology. The discussion will not be comprehensive, either in its examination of technologies or in its description of extension approaches. The objective is simply to ensure that the concept of LEIT is sufficiently well illustrated so that there is a clear basis for the analysis in the remainder of the book.

The outline of types of technology follows the discussion in Chapter 1, which focused our attention on technologies that are based on local resources and/or reduce or eliminate reliance upon external (non-renewable) inputs. Although such technologies may be introduced as part of a more comprehensive vision of sustainable development or rural regeneration, our interest is in the technologies themselves and the methods available for introducing them. This pragmatic approach means that we are willing to examine examples of the technologies and their diffusion in a wide range of circumstances and contexts.

The review of technologies and methods will be almost exclusively limited to examples from the South, although there are many relevant experiences to be found in industrialized countries. The technologies may be promoted by external agents (and taken up by farmers) for a variety of reasons, including short-term productivity advantages, long-term resource development or environmental protection. They may substitute for purchased inputs (manure in place of fertilizer) or play a role for which no alternative external resources are relevant (terracing). The examples in this discussion are limited to crop management techniques and do not involve varietal selection or plant breeding.

The review will also concentrate on examples in which LEIT is introduced, rather than being part of traditional farming systems, although the distinction is often difficult to make, especially given the fact that much of LEIT is based on local practice. But our interest is to understand the trajectory of LEIT projects that promote the introduction of novel methods. So, for instance, we shall spend little time in this chapter discussing indigenous soil management technologies, even though they form the basis of much useful LEIT (Wilken, 1987; Reij et al, 1996; Butterworth et al, 2003; Widgren and Sutton, 2004). However, the division between local and introduced is somewhat artificial, and many LEIT projects correctly devote considerable effort to understanding and improving upon local crop management techniques. For example, Adégbidi et al (2004) report from Benin on the recognition of the importance of farmers' continued use of a range of traditional soil and water conservation measures in the midst of previous unsuccessful government campaigns to introduce alternative measures.

There is also another side to the question of 'local' technology, and tracing its origins often underlines the importance of exchange, contact and adaptation in technology innovation. Buckles (1995) describes the recent history of velvet bean (*Mucuna pruriens*), one of the most popular cover crops promoted in Central America (and elsewhere). The crop, of Asian origin, was being grown in the southern US as a forage and green manure as early as 1915 and was introduced into Central America as a forage crop by the United Fruit Company during the 1920s to feed the mules used on its plantations. Some indigenous highland Guatemalans working on the plantations carried velvet bean with them when they migrated to lowland valleys and used it as a fodder and green manure. Some of these farmers subsequently brought velvet bean to coastal Honduras, where, during the 1980s, it spread rapidly to the point where it was part of the hillside maize management strategies of nearly two-thirds of the small-scale farmers in certain areas. It was subsequently promoted in other areas by a range of development projects. Similarly, a rice farmer in Sierra Leone once told a puzzled researcher about '*Mende* fertilizer', which the farmer recognized as a weed that may be nurtured because its presence in fallow land contributes to soil fertility. Rather than being a native plant, it turned out to be a species introduced by the colonial authorities in 1938 for green manure experiments (Richards, 1995, cited in Anderson et al, 2001).

Not only is the local/introduced dichotomy difficult to maintain, but any attempt at a functional classification of LEIT faces many problems. This is partly because many examples of LEIT perform various functions (an inter-crop may reduce insect attack and contribute to weed suppression) and partly because individual examples of LEIT are often part of complex and integrated crop management strategies in farming systems that are themselves evolving and encountering new challenges. The following introduction to some important cases of LEIT is divided very crudely and, at times, arbitrarily into examples of: soil and water management; soil fertility enhancement; crop establishment; and weed, insect and pathogen control. This does not pretend to be a comprehensive review, but is only meant to illustrate some major examples of LEIT, many of which will be explored later in the book. The discussion does not attempt any type of technical assessment, but does try to point out major implications for the management of labour and information, themes that are of particular importance for the subsequent analysis.

Similarly, there is no immediately obvious classification system for the techniques that are used to develop and promote LEIT. The final part of the chapter merely outlines some of the major strategies that are important for LEIT projects and briefly describes a few of the most prominent techniques.

Soil and water management

There are a variety of physical or biological interventions that can be used to protect the soil from erosion and to channel water runoff or conserve soil moisture. These are resource-conserving technologies, many of which have no 'external input' substitutes, but they are often linked to the introduction of other types of LEIT. Major mechanical methods include the construction of terraces, bunds or channels. Soil preparation techniques can also be modified to prevent erosion and conserve moisture, as in contour ploughing. Vegetative material can be arranged along the contour and various crops can be planted to stabilize hillsides, as in the practice of contour hedgerows. Finally, there are techniques for reducing or eliminating tillage; these often include the use of mulches or cover crops that help to suppress weeds and conserve moisture.

Physical structures

One of the major techniques for soil conservation in hillside farming is the construction of various types of terrace. Such technology is of ancient origin, and its large labour requirements have often been met by community organization or state coercion. However, there are also techniques that are more amenable to local initiative. One type of terrace that has been used widely in Kenya is known as *fanya juu* and apparently originates in farmers' practices in the east of the country (Tiffen et al, 1994). It consists of digging a narrow trench along the contour and throwing the soil uphill, allowing it to form a ridge. If the ridges are stabilized by planting live barriers, small bench terraces are eventually formed through erosion and deposition of soil. Although this requires considerable labour, it is less demanding than the digging and construction of conventional terraces. Locally derived techniques such as this are not always widely applicable, however. When *fanya juu* methods were promoted in western Rwanda, where soils are thinner and less fertile than in Kenya, the terraces contributed to land degradation as infertile soil horizons were brought closer to the surface (Johnson and Lewis, 1995).

Terracing of various types also helps to slow water runoff and direct its movement, and additional specialized structures may also be useful for this purpose. Cut-off drains may be dug to conduct water away from the field, which often involves coordination among a number of neighbouring farms. Where gullies have already formed, small check dams may be constructed to prevent further expansion, and vegetation may be encouraged to stabilize and help recover the land. A less extensive technique is the construction of infiltration ditches in fields to catch and hold water runoff, although these still require considerable labour and remove a portion of the field from cultivation.

Where stones are available, these can be transported and arranged to form terraces. Substantial stone terracing is sometimes found in hillside agriculture, and simpler stone bunds are effective where the slope of fields is less severe. One of the most widespread recent successes in soil and water conservation is based on the construction of stone bunds along contour lines in farmers' fields in Burkina Faso (Atampugre, 1993), one of the major contributions of the technology being moisture conservation in this area of dryland agriculture. The technique is an elaboration of local practices and formed the basis of large-scale projects in the region.

Contour planting

Whether hillside fields are prepared by hoe or plough, farmers can adopt contour planting to enhance soil and water conservation. Many development projects train farmers in the use of the A-frame (or other easily manageable techniques) to lay out ploughed or hoed ridges on the contour.

An interesting innovation known as in-row tillage has been developed by agricultural projects in Central America (Bunch, 1999). The technique involves ploughing or hoeing land only in the row where the crops are planted, leaving the intermediate area untouched. If the practice is carried out over several years, mini-terraces gradually form on their own, or additional work in the first year can develop such terraces. The process is enhanced by the addition of crop residues and other waste to the intermediate area (see Chapter 5).

An elaboration of contour ridging is the practice of tied ridges. Various methods exist for improving upon the ridges traditionally created by plough or hoe in order to better conserve moisture in semi-arid agriculture. A considerable amount of effort was devoted to promoting tied ridging in the Sahel, where depressions were created by constructing small ridges perpendicular to (and slightly lower than) the planting ridges every 1 to 2 metres (m) (Nagy et al, 1988). These depressions collected rainwater, enhanced infiltration and made the application of mineral fertilizers more productive. However, farmers found their construction demanding too much labour (especially at a time when other tasks were pressing). Various attempts have been made to develop mechanical (animal-drawn) tied ridgers; but they have yet to find widespread use. Although tied ridging has not been widely adopted in response to formal promotion in the Sahel, there are many areas where some variation of the practice is traditionally used; for instance, Adégbidi et al (2004) describe the continued use of tied ridges on hillside fields in northern Benin.

Biological innovations

In addition to physical constructions or adjustments to soil preparation, soil and water conservation can also be enhanced by the use of vegetation. Trash lines can be formed by placing cut weeds, crop stover, twigs and other materials along a contour, and allowing the resulting barrier to gradually form a bund as soil collects against it. Mercado et al (2001) describe how hillside farmers in the Philippines experimented with placing crop residues along contours to form 'trash bunds' that allowed native grasses and weeds to grow and form natural vegetative strips. These proved popular among farmers who had previously experimented with hedgerow intercropping but found it too labour demanding.

Hedgerow intercropping for erosion control consists of sowing particular species along the contour for the purpose of forming a live barrier. The hedges constitute 'biological terraces' through the accumulation of soil on the upslope side of the hedge, although such soil movement often manifests itself in fertility gradients, with higher yields recorded on the lower parts of the terrace (Cooper et al, 1996). Where soil erosion is particularly high, the effectiveness of the hedges is increased if prunings are used as mulch. However, farmers often prefer to plant trees or shrubs of economic value as hedges, limiting the possibility of mulching. Another alternative is to use tree–shrub combinations to stabilize and help develop existing conservation structures. Where possible, the species should have economic value (e.g. for fuel, fruit or fodder), although the choice will also have to take into consideration the competition for nutrients and moisture with the principal crop in the field (Cooper et al, 1996). Live barrier options thus present a set of trade-offs for farmers to consider. On the one hand, they usually require less maintenance than physical barriers and may provide additional products to the farm; on the other hand, they may occupy additional space, require protection from grazing animals and may compete with the principal crops for moisture or soil nutrients.

One species used for live barriers that has drawn considerable attention is vetiver grass, a plant of Asian origin that develops an exceptionally deep root system and, when planted on contours, forms a dense vegetative barrier that retards runoff and develops bunds (NRC, 1993). It has been used for centuries in India for various purposes, including stabilizing the banks of canals. Vetiver rose to prominence after it was successfully used in Fiji to stabilize hillsides where commercial sugar cane was to be planted. The World Bank subsequently promoted the use of vetiver in a range of settings and although it has lost some of its 'miracle crop' status, it still features in many soil erosion-control projects. When successfully established it can be a very effective and hardy living barrier; but it is not appropriate for all growing conditions. For instance, research in Honduras found that vetiver barriers were not particularly effective at controlling soil loss, nor did they contribute to additional maize yields on hillside plots (Hellin and Schrader, 2003). Its initial establishment requires a nursery source for planting material and it often takes some time and management to create a viable hedge. Other grasses related to vetiver, such as lemongrass, are also commonly used in living barriers.

One of the problems with vetiver is that it has few uses, and farmers complain about the sacrifice. Another grass species that can offer a profitable alternative is Napier grass (or elephant grass), which is an excellent fodder crop. Although it does not form a particularly dense hedge, and its shallow, spreading roots compete with those of nearby crops, it offers a more economically attractive option than many alternatives. It has been successfully used in soil and water conservation efforts in Kenya (see Chapter 6) and Guatemala (Bunch, 2002).

Conservation tillage

Because traditional tillage operations often contribute to soil and moisture loss, there are a range of techniques for reducing or eliminating tillage operations (referred to by a variety of terms, including zero tillage and conservation tillage). Traditional slash-and-burn agriculture is one way to avoid tillage, but this depends upon the availability of large amounts of land for shifting cultivation. Tillage has been traditionally performed to prepare an adequate seedbed and to control weeds, so attempts to reduce tillage (and thereby control soil and mois-ture loss and save land preparation costs) must offer adequate alternatives. When tillage operations are reduced or eliminated, farmers need a method for crop establishment, and this may vary from a simple planting stick, to an ox-drawn ripper tine (that opens a narrow slit in the soil where the seed is deposited), to various types of seed drills. Most reduced tillage systems include some type of mulch to help protect the soil, control weeds and conserve mois-ture. The mulch may comprise only crop residues or may include a cover crop. The mulch by itself is usually not sufficient to control weeds completely and many reduced tillage systems are dependent upon some use of herbicide, a fact that some proponents of sustainable agriculture find difficult to accept but oth-ers recognize as a necessary compromise. Herbicides are not a requirement for conservation tillage, however, and systems have been devised that rely upon cover crops for weed control (e.g. in subtropical southern Brazil).

The vast majority of successful conservation tillage reported in the litera-ture is found in Latin America, and although its practitioners include small farmers, it is generally a component of commercial, mechanized agriculture with a relatively high dependence upon external inputs (Landers, 2001; Ekboir, 2002). The conversion to conservation tillage in Latin America (and in industrialized countries such as the US) has involved a great deal of adaptive research and extension since farmers have to learn a number of new principles and techniques and new equipment needs to be designed and tested. The participation of existing farmer cooperatives in Brazil, and the formation of Friends of the Land clubs that allowed interested farmers to exchange experi-ences, was important in creating a critical mass of support for conservation tillage in Brazil (Landers, 2001).

Mulches and cover crops have a wide range of applications beyond contributing to reduced tillage systems. Farmers in Burkina Faso are increas-ingly mulching their fields by cutting grass and transporting it to the field, where its decomposition adds nutrients to the soil and attracts termites whose activity improves soil quality and workability (Slingerland and Masdewel, 1996). Cover crops may be grown in association with a field crop but often are left to grow after the main cropping season. They may serve simply to protect the soil from erosion, but are often used to enhance soil fertility (as a green manure); in the latter case, the objective is to let the cover crop grow as long as possible so that the maximum amount of organic matter can be incorporated within the soil.

The term cover crop is used in a variety of ways. Anderson et al (2001) distinguish the functions of cover crops (the biophysical processes in which cover crops are involved) from their purposes (the uses to which they are put by farmers). The distinction is important because it emphasizes the fact that farmers often focus on a subset of a cover crop's agronomic characteristics in making management decisions. Their functional definition of a cover crop is 'a live soil surface cover ... which performs one or more functions such as organic matter production, nutrient recycling, allelopathy and/or competitive growth, soil cover, nitrogen fixation and insect suppression, while producing forage and grain' (Anderson et al, 2001, p109). The challenge in introducing cover crops is matching a particular species' set of functions to local farmers' purposes. For instance, in promoting the cover crop velvet bean in Benin, researchers emphasized the function of nitrogen fixation in order to encourage better soil fertility management; but farmers based their decisions more on their purpose of identifying an effective weed suppression agent (Versteeg et al, 1998).

Because cover crops have a range of functions, the decision to utilize them usually involves considering a series of trade-offs, and farmers need as much information as possible about the functions of a particular species. For instance, a cover crop may contribute to enhancing soil fertility; but this may imply sacrificing a second season of crop cultivation to allow the cover crop to develop. Farmers need experience to see how the multiple functions of a cover crop (e.g. fertility enhancement, moisture retention and weed suppression) may be relevant to their particular purposes. There are also trade-offs in terms of labour, and some cover crop management systems imply a long-term increase in labour requirements (e.g. where the cover crop makes harvesting or land preparation more difficult), while other systems lead to reduced labour inputs (e.g. where the cover crop substitutes for some manual weed control or makes the soil easier to prepare for planting). Additional considerations include possible difficulties (e.g. in acquiring or maintaining seed) and advantages (e.g. in using the cover crop for forage).

Soil fertility enhancement

A considerable amount of effort has been devoted to developing LEIT in order to help replace or complement the use of mineral fertilizers. The interventions include manures and composts, as well as crop species that can be used in biomass transfer, alley cropping or as green manures. Although it is widely recognized that over-reliance on mineral fertilizers is both economically and environmentally unwise, there are many challenges to developing effective regimes for providing organic soil fertility amendments. Many types of manure have fairly low nutrient concentrations and their production, of course, depends upon the availability of grazing or fodder resources. Crop residues are also often relatively low in nutrients and their use slows, but does not reverse, the depletion of soil nutrients from the crop harvest. Leguminous crops offer many possibilities; but their productivity may be limited by their growth habit or by lack of soil phosphorus or appropriate strains of rhizobium. Leguminous and other species grown specifically as soil amendments usually imply learning about the new crops and their management, as well as investing additional labour, and may also involve the sacrifice of some crop land.

Giller (2002) discusses the importance of making better use of organic soil fertility resources, supplemented with mineral fertilizers. In the African context, where soil fertility management is a key to agricultural development, he points out that there are fewer interactions (synergies) between organic and mineral sources of soil nutrients than many would hope. Farmers need information on how to combine the use of organic fertilizers (with their multiple benefits, but considerable management challenges) with mineral fertilizers that are more limited in their range of benefits, but amenable to more precise control (e.g. in timing and placement).

Manures and composts

A traditional LEIT for soil fertility enhancement is the use of animal manure. Manure is a particularly effective way of converting low-quality plant material to a more efficient source of nutrients. However, because most animal manures have quite low nutrient contents (as a rule of thumb, smaller animals produce manure of higher nutrient concentration), their benefits may be derived as much from their contribution to soil physical structure as from nutrient supply *per se* (Graves et al, 2004). The incentives for increasing the use of manure as a soil amendment are higher where population density is high, animals are confined and fields are near enough to the homestead to make transport costs acceptable. Manures may be mixed with crop residues or other vegetative matter or composted.

An important limiting factor for manure use is animal ownership. A study in northern Tanzania looked in detail at soil fertility resources available to different classes of household. Those households classified as resource-rich had nearly 7 times the amount of manure available as medium-resource households and nearly 50 times that available to the poorest households. Although landownership was not highly skewed, the fact that wealthier households rented in (and poorer households rented out) meant that the soil fertility resources of the wealthy households were applied to 5 times the area cultivated by the median households and 11 times that farmed by the poorest (Piters and Ndakidemi, 2000).

Another effective example of LEIT for soil fertility is composting, a traditional practice that allows farmers to enhance and make better use of resources available to them. Compost provides a much more effective source of nutrients than most of the unprocessed organic matter that farmers may apply to their fields. However, composting has received relatively little attention in low-input agriculture projects, perhaps in part because of its association with former top-down extension campaigns. Other explanations for the limited spread of the technique among resource-poor farmers are the relatively large amounts of biomass that are required, competition with other uses for that biomass (fuel and fodder), and the labour necessary to prepare and manage the compost pit and transport the product (Graves et al, 2004). In part because of its limited availability, compost tends to be applied in restricted areas for high-value crops.

One example of composting that appears with increasing frequency in the literature on sustainable agriculture is vermicomposting. The production of vermicompost is becoming a popular activity for non-governmental organizations (NGOs) in India and elsewhere (e.g. Butterworth et al, 2003). Vermicompost is obtained by raising certain types of earthworm that consume surface litter (rather than live in the soil); the worms must generally be specially reared and supplied. The farmer needs to construct a suitable composting bed, preferably with brick or concrete sides, to contain the worms. The bed is filled with layers of leaves or other trash, decomposed manure and other organic matter, with the finer material near the top. The bed may be covered with plastic or banana leaves. Once the worms are introduced the bed must be watered carefully and the worms must be protected from various predators. The farmer periodically adds organic matter to the bed and the vermicompost is ready after three to six months. The product is a high-quality, fine-textured compost; but its production requires careful management.

New species for soil fertility enhancement

There are an increasing number of examples of efforts to introduce new species to farming systems that can contribute to soil fertility enhancement. The management of these innovations includes biomass transfer, the formerly highly prominent practice of alley cropping, and various types of green manures.

Biomass transfer involves growing a crop in a separate field that can be cut and carried to provide organic matter. Farmers have traditionally cut vegetation and carried it to their crop fields; but a number of additional techniques have been devised for planting crops for this purpose on otherwise unused land. The effectiveness of the strategy depends upon finding species whose leaf litter has an acceptable nutrient content, which decompose relatively rapidly and are available at the appropriate time (Cooper et al, 1996). Because the nitrogen content of leaf litter is generally low, it is also necessary to contemplate the production of a significant amount of biomass if the technique is to be worthwhile. Although most of the species that have been used are leguminous, recent work has identified potential for *Tithonia diversifolia* (Mexican sunflower), which is able to extract significant quantities of phosphorus from the soil as well as providing an acceptable nitrogen content (Graves et al, 2004). The biomass transfer techniques can bring additional nutrients to crop fields; but the maintenance of the biomass bank and the labour involved in cutting and carrying the material must be taken into account.

Alley cropping is a system of agroforestry management that was designed as an alternative to shifting cultivation systems in tropical Africa. The system is based on planting rows of multipurpose trees alternately with rows of crops. The tree species are leguminous and their cuttings provide mulch and green manure for the field, as well as fodder for animals. The stems can provide material for fencing or firewood. The shade of the intercrop also reduces weed competition. Various species have been used, although the most common are *Leucaena leucocephala* and *Gliricidia sepium* (originally from Central America). Alley cropping evoked a great deal of hope in the early years of its development during the late 1970s (e.g. Harrison, 1987) but has since fallen from favour. One of the problems was the fact that the system was best adapted to higher rainfall regions; in drier areas the leguminous intercrop competed with the grain crop for moisture and nutrients (Graves et al, 2004). But perhaps the most significant factor limiting the acceptability of alley cropping was the higher labour requirements (for nurseries, establishment, pruning and clearing alleys) and the fact that many of these labour demands had to be met at a time when other farm operations were pressing (Carter, 1995).

Various crops can be introduced as green manures to enhance soil fertility. Farmers may allow the crop to grow for as long as possible and then simply incorporate it within the soil at the next land preparation, or they may periodically cut or harvest green material (as in alley cropping). For instance, *Tephrosia vogelii* is a leguminous shrub planted along with maize in Southern Africa. It matures more slowly than the maize and does not compete with it, but requires careful weeding in the early stages. If the field is replanted the following year, the standing *Tephrosia* is cut at the time of land preparation and allowed to dry in the field. The leaves are then shaken off and left as mulch or incorporated within the soil. If the field can be left fallow for a year or more, the resulting growth makes a larger contribution to soil fertility.

Leguminous crops such as *Tephrosia vogelii*, *Sesbania sesban* and *Crotalaria grahamiana* are also promoted as improved fallows. They can be planted with the main crop and develop slowly at first. They are then left in the field after harvest and allowed to grow for at least another season. Farmers then harvest these improved fallows before the subsequent planting season, removing the stalks (which can be used as firewood) and incorporating the leaves, stems and other debris to provide nutrients for the next grain crop. Such a system helps to address the breakdown of traditional fallow systems that were previously used to regenerate soil fertility. The system is most applicable in areas where there is sufficient land to allow fallowing; but there is evidence that even in some areas of Africa where land is quite scarce farmers leave part of their holdings fallow in the hope of some fertility regeneration (Cooper et al, 1996). The production of an improved fallow requires learning about the management of a new species and additional labour in establishing the plot and doing initial weeding; but these labour requirements are not as demanding as those of alley cropping.

Velvet bean (*Mucuna pruriens*) has had a particularly interesting career as a green manure in Central America; it is a prominent element in many agricultural development efforts, but it also spread autonomously in parts of coastal Honduras through farmer-to-farmer diffusion (Buckles et al, 1998). Although *Mucuna* offers several advantages, it is valued principally as a soil amendment in Central America, where farmers refer to it as the 'fertilizer bean'. On the north coast of Honduras, farmers can plant in two seasons. They typically establish *Mucuna* in the second (winter) season by planting into a maize field and allowing the *Mucuna* to mature after the maize harvest. The field is left during the summer season as a short fallow; the *Mucuna* matures, senesces and eventually reseeds itself. The following winter season the field is prepared simply by slashing the *Mucuna* to the ground and planting maize into the mulch. The field requires only light weeding and little or no fertilizer, and the system allows continuous cropping. In areas of lower rainfall (such as southern Mexico), farmers using *Mucuna* may need to replant it each season, and in areas of land shortage, where farmers get two maize harvests in a year, the *Mucuna* may be inter-planted and then slashed at the end of the same season.

Crop establishment

The examples described in this section illustrate the complex, multipurpose nature of many types of LEIT. Although these are all depicted as techniques for crop establishment, they perform various functions in the farming system. Two of the techniques are recent innovations that have received considerable attention in the literature: planting pits and the system for rice intensification. The third example involves a brief discussion of the traditional practice of intercropping.

Planting pits

There are a number of traditional techniques found in Africa where farmers construct small pits or basins and plant directly in the depression or on its side (in the case of the *ngoro* system in Tanzania, described by Malley et al, 2004, farmers plant on the top of the ridges that form a lattice of pits, with an effect similar to that of tied ridges). In some areas in the Sahel, farmers have traditionally used planting pits (*zai*) on a small scale to rehabilitate degraded land, particularly hard or gravely soils with little organic matter. The technique has been elaborated upon and promoted by a number of projects in Mali, Burkina Faso and Niger (Hassan, 1996; Wedum et al, 1996; Kaboré and Reij, 2004). In its simplest form it involves digging a series of small pits and, perhaps, placing a small amount of manure or compost in each. The pits are used as planting stations where fertility and moisture are concentrated. The planting pit methodology promoted more recently involves digging somewhat larger pits (about 20cm to 30cm in diameter and 10cm to 20cm in depth) about 1m (or less) apart throughout a field. The pits are filled with as much manure or other organic matter as is available. If the pits are dug in fields subject to runoff, they may be complemented by half-moon earth bunds downslope to help collect water.

Digging the *zaï* demands a great deal of labour; one study estimates 40 person days per hectare (Hassan, 1996) and another estimates two to three times that (Wedum et al, 1996). The planting pits must usually be reformed (in between former pits) after several years. There seems little doubt that the technology can make a significant contribution to improving yields by conserving moisture and concentrating plant nutrients. It also improves soil structure and protects young plants from winds. The planting pit technology cannot be combined with ploughing, however, and is not suitable for all soils. One of the limitations is the amount of manure available. Some experimental work has shown that mineral fertilizers used with planting pits can increase yields further; but the economic return may not be acceptable in years of exceptionally low rainfall (Hassan, 1996). The technique has recently featured in a nationwide campaign in Zambia, under the somewhat confusing title of 'conservation tillage', as an alternative way of managing traditional food grain fields where productivity has declined because of lower use of mineral fertilizers (after subsidies were removed) and the development of plough pans from mechanical soil preparation (Haggblade and Tembo, 2003).

The system of rice intensification (SRI)

The system of rice intensification (SRI) is a relatively recent innovation and one of the most controversial examples of LEIT (Stoop et al, 2002; Doberman, 2004). It was pioneered by a French priest working with farmers in Madagascar and is a relatively complex system of practices rather than a single intervention. It is based on a planting method that upsets several conventions for irrigated rice. Under SRI, the rice is transplanted much earlier than normal (one to two weeks after germination), the transplanting is done rapidly and with great care so as not to damage the roots, and the seedlings are individually planted at a wide spacing, resulting in a much lower initial plant stand than with normal transplanted rice. Another major difference is that the soil is kept well drained, at least until the flowering stage; irrigation (if available) is reduced and managed so that there are short alternate periods of wetting and drying.

Exceptionally high yields have been reported for some experiences with SRI, although these results are the product of quite careful management. There are sacrifices involved in removing rice from a flooded environment (such as increased weed competition), but there are apparently physiological mechanisms that are brought into play by the alternate wetting and drying and the wide plant spacing that can contribute to increased yields (e.g. by promoting much more vigorous root growth and increased tillering) (Stoop et al, 2002). Critics have questioned some of the reported yields and have begun to point to agronomic limitations on the replicability of the technique (Doberman, 2004). SRI certainly requires additional labour during the learning stages, although farmers may be able to reduce these demands with additional experience. The radical shift in the growing environment for rice brought about by SRI also implies a reconsideration of complementary crop management practices. Most versions of SRI include a significant reduction in mineral fertilizer (and its substitution by organic fertilizers) and mechanical rather than chemical weed control. Most SRI proponents have placed themselves firmly in the 'low-input' camp, which serves to raise the profile of the debate but may lower the incentives for a pragmatic approach to exploring the technology's potential.

Intercropping

Finally, any discussion of crop establishment techniques should include a few words about the general practice of intercropping, which is often associated with low-input agriculture. Farmers have long exploited the advantages of intercropping. Individual crop yields may suffer somewhat because of competition, but the net gains in production are often much greater than what could be obtained from dividing the equivalent area into separate sole-cropped fields. Intercropping usually makes good use of the different growth habits of the various crops so that there is complementarity in the use of time (different maturity periods), space (different root or canopy arrangements) or crop physiology (Francis, 1986; Altieri, 1995).

Modern agriculture is often associated with the demise of intercropping practices since commercial incentives favour particular crops and external inputs interfere with, or obviate the need for, another crop in the field. The intercropped field has attained mythical qualities in the eyes of some campaigners for traditional agriculture (e.g. Shiva, 1993); but an empirical approach is required. Intercropping is not always the most appropriate answer, even under low-input conditions (e.g. Lightfoot et al, 1987) and there are cases where attempts to convert a traditional monocrop to an intercrop have not proved successful (de Jager, 1991). Where intercropping is the norm, there may be a very wide range of crops and mixtures employed, and farmers' experience and experimentation usually ensure that the intercropping practices strike a reasonable balance among factors such as seasonal production needs, labour availability and risk aversion. It is thus relatively unlikely that researchers will be able to offer any improvements on the local practices; but the introduction of new species, such as green manures, can offer significant benefits and take advantage of farmers' familiarity with intercropping techniques.

Controlling weeds, insects and pathogens

Weed control

Until recently, all weed control was accomplished with local resources, primarily the hoe or the plough or, in the case of slash-and-burn systems, the use of fire. At present, LEIT weed control options are contrasted with the use of herbicide, which has spread rapidly even in small farm and subsistence agriculture (Naylor, 1994; Sain and Barreto, 1996). Several of the technologies described earlier, such as cover crops, alley crops and mulches, make important contributions to weed control. If reduced tillage options are to control weeds without resorting to herbicides, they must usually include mulches or cover crops. We have seen that in some cases cover crops that are promoted primarily as soil fertility amendments are adopted by farmers largely because of their weed suppression capabilities. Intercropping can also contribute to weed control (as when a grain crop is grown in association with another crop whose growth habit provides good ground cover), although it is usually difficult to promote an intercrop solely on the basis of its contribution to weed control.

Reducing pesticide use

A much more prominent target for LEIT is the problem of the widespread use of synthetic insecticides, fungicides and other chemicals for controlling pests and pathogens. The misuse of cheap broad-spectrum insecticides has led to serious health problems among farmers, has been responsible for significant environmental damage (upsetting local ecologies and poisoning groundwater), and has caused economic losses (Rola and Pingali, 1993). There are a number of alternatives designed to reduce or eliminate pesticide use. The following discussion includes a few specific examples and concludes with an outline of the generic term, integrated pest management (IPM), and its applications.

Altieri (1995) provides many examples of how crop rotations and intercropping can be used to control insect pests. There is good evidence that intercropping practices can serve to lower the incidence of insect pests by attracting more natural enemies to the field and/or by interfering with the pest's ability to attack its target. It is possible to choose intercrops or companion crops to respond to threats from specific pests; but the information required to develop such precise management requires considerable research. Altieri (1995) provides an exceptionally complex example from the North of how field crop management might be organized to control a pest (the onion maggot) on farms in Michigan. The onions are planted in long, narrow strips, in close proximity to weedy borders (which provide nectar for a maggot parasite) and pasture (which fosters earthworms that help to attract a predator of the maggot). The weedy borders are left un-mowed in order to trap flies affected by a common disease whose spores can then infect healthy flies. Radishes are planted nearby to provide a host for another predator of the pest, and an early crop of onions (planted in close proximity to the radishes) serves as a trap crop for the maggots. Needless to say, this type of management is the exception rather than the rule, even in commercial agricultural systems that can afford the research to develop this information.

A method for insect control in maize that has attracted recent attention is the 'push–pull' strategy for controlling stem borers in maize in Africa (Khan et al, 2001). In experiments in Kenya, farmers planted a 1m-wide perimeter of Napier grass or Sudan grass around their maize plots. These grasses serve as a trap crop as the stem borers prefer to colonize them rather than the maize. The maize fields themselves are intercropped with molasses grass or the leguminous fodder crop *Desmodium*, a species that repels ovipositing stem borers. The experiments have found that the *Desmodium* intercrop offers the very important additional advantage of significantly reducing infestation by the parasitic weed striga, which is a major cause of yield loss (the molasses grass offers its own advantages, as it attracts a natural enemy of the stem borer). The 'push–pull' system is an attractive example of LEIT that has found its way into the literature on sustainable agriculture (e.g. Pretty, 2002), although the extent of uptake by farmers is not yet clear. The method's considerable advantages of insect and weed control are balanced by the fact that it asks small farmers to dedicate a significant amount of land (and labour) to establish intercrops and borders; in addition, the fodder crops are of value principally to farmers who own cattle, and desmodium seed is expensive.

Integrated pest management (IPM)

Many of the LEIT efforts in pest control are referred to as integrated pest management (IPM). Although it is possible to encounter a wide range of IPM activities, the term has no simple definition. It is probably easiest to delimit IPM by stating what it is *not*: the preventive, calendar-based application of insecticides or other pesticides. Most examples of IPM share a belief in the use of a wide range of methods to maintain pest damage below an economic threshold, the necessity of an iterative approach that considers trade-offs and compromises, and attention to understanding and promoting ecological control mechanisms (Jeger, 2000). The knowledge and resources that are brought to bear in IPM may include chemical pesticides, although for some practitioners IPM is synonymous with chemical-free or organic production.

The term emerged during the early 1970s, based on a great deal of earlier work on biological pest control followed by what came to be called integrated pest control, which was a compromise between those working on biological control and entomologists favouring chemical control (Palladino, 1996). Many early examples of IPM concentrated on the concept of economic threshold, where farmers were taught to estimate pest and predator populations (through various counting methods) and to make pest control decisions accordingly. Most of these methods proved to be too time-consuming and complex for farmers to master (Morse and Buhler, 1997). Other efforts relied heavily upon community coordination (e.g. for synchronous planting or setting traps); but these strategies were often difficult to manage.

Much of the early experience in testing practical IPM solutions came in the context of Green Revolution irrigated rice fields in Asia, and it was here that one of the most successful examples of IPM emerged. Workers in Indonesia during the late 1980s began experimenting with the concept of a farmer field school (FFS) in which farmers would become acquainted with agro-ecological principles in a social learning context. These principles, by illustrating examples of the natural control already taking place in the fields, would give farmers confidence to reduce their use of insecticides (Röling and van de Fliert, 1994; Kenmore, 1996). The FFS also taught other pest control techniques (such as management practices and natural pesticides); but it would appear that the largest impact of this type of training was simply encouraging farmers to think carefully before reaching for the sprayer, thereby significantly cutting the amount of insecticide being used (see Chapter 7). There were several factors that contributed to this success. First, rice farmers' use of insecticides had increased rapidly at the time of the Green Revolution, in part because the technology allowed more frequent planting and some of the new varieties were more susceptible to pest attack. Government programmes were based on a 'package' approach to extension, and pesticides were often a part of the package, whether they were really needed or not; commercial pressures also encouraged farmers to use pesticides. As farmers began to understand that some pesticide use could actually be counterproductive (by destroying natural enemies), and as further generations of modern rice varieties proved to be less susceptible to attack, opportunities emerged to break what had become, in some cases, a dependence upon cheap, broad-spectrum insecticides. In addition, the irrigated rice system is perhaps unique in that it is a centuries-old ecosystem which has evolved exceptionally complex and effective regulating mechanisms (Settle et al, 1996). The promotion of insecticides as part of production campaigns had upset this balance; but it was possible to help restore ecological mechanisms by encouraging rice farmers to reduce their insecticide use.

There are many other programmes and projects aimed at introducing IPM in other crops and environments, and they meet with variable success (Jeger, 2000; Way and van Emden, 2000). Most cropping systems are of more recent origin than irrigated rice, and natural control mechanisms are often less well developed. Thus, in general, IPM (for example, in crops such as cotton or vegetables) requires learning not only the importance of lowering or eliminating insecticide use, but also learning about complementary techniques and inputs that can replace the chemicals. Some IPM projects focus on a few specific techniques and place considerable weight on providing farmers with agro-ecological principles that will help them make more informed choices. Other projects denominated as IPM may simply work to reduce insecticide use or substitute less dangerous products. Still other IPM efforts feature a mix of technologies (such as resistant varieties, bio-pesticides, alternative management techniques and light traps) that are often offered as individual options, with little emphasis on the 'integrated' part of the strategy. In some cases farmers may be introduced to a series of new technologies without adequate testing under local conditions or with unrealistic aspirations regarding their capacity to do their own adaptation (Orr et al, 2001). The various IPM strategies offer important opportunities for safer and more sustainable pest control; but few of them represent the simple substitution of one practice for another and most require farmers to continually update their skills and their knowledge of new techniques and to be ready to adapt to changing conditions.

Methods for generating and promoting LEIT

Although there is no strict correspondence between those technologies described as LEIT and particular methods of promotion or extension, the interest in LEIT has been associated with a number of innovations for interacting with farmers in the development and diffusion of new technology. These innovations are based on some general principles (including the promotion of indigenous knowledge systems, the facilitation of farmer-to-farmer interchange and support for farmer experimentation), as well as including a number of formal methods, several of which are introduced below.

General approaches

The development of productive LEIT usually depends to at least some extent upon farmers' capacities to innovate, their willingness to adapt general principles of resource management, or their elaboration of management techniques for new species such as green manures. This implies respect for local knowledge systems and support for farmer innovation capacity. This is not the place to review the considerable literature on indigenous technical knowledge (ITK); but there are practical examples of how the strengths and weaknesses of ITK must be understood to promote the introduction of LEIT. For instance, Bentley (1989) discusses the implications of farmers' knowledge about pest behaviour for the development of IPM programmes. Local farmer innovators are the source of many examples of novel crop management practices, as well as being energetic collaborators in testing new ideas. There are some initiatives that focus specifically on the promotion of such farmer innovators and the facilitation of their experiments and communication (e.g. Reij and Waters-Bayer, 2001).

When LEIT involves the introduction of fairly complex technologies that require local adaptation, farmers often profit from 'cross-visits', opportunities to observe such technologies in practice in neighbouring (or further removed) communities and to discuss techniques and challenges with those farmers who have more experience. The *Campesino a Campesino* movement in Nicaragua, promoted by the National Union of Agricultural Producers during the Sandanista regime, aims to help small farmers acquire environmentally sound production techniques. It makes considerable use of exchange visits, where members of one community visit those in another to learn about innovations. The movement has found that farmers are particularly effective at communicating their experiences to their counterparts (Anderson et al, 2001; Holt-Giménez, 2001). Similarly, Sahelian farmers who are being introduced to conservation technologies such as planting pits, which involve a considerable labour investment, have profited from visits to other areas where the technology is in place in order to see what a difference it can make to crop production (e.g. Kaboré and Reij, 2004).

Because LEIT involves the adaptation of general principles, using local resources and knowledge, top-down extension techniques are often inappropriate. In addition, LEIT is often part of broader efforts aimed at empowering farmers to play a larger role in the generation and control of technology, so alternatives to standard extension strategies must often be sought. However, there are many examples where LEIT is introduced through conventional extension techniques. There are many instances where soil and water conservation techniques have been promoted in a top-down and, at times, even authoritarian manner, although such strategies have often been ineffective (Stocking, 1996; Hellin and Schrader, 2003). Various IPM techniques and inputs may be promoted through fairly conventional extension techniques (Orr et al, 2001; Tripp and Ali, 2001), and considerable success has been registered in lowering farmers' insecticide use in rice through the promotion of simple rules on the time of spraying (Heong et al, 1998). Global 2000, which has traditionally concentrated on seed-fertilizer technology promoted through large demonstration plots, successfully included promotion of the cover crop *Mucuna pruriens* in some of its West African programmes (Manyong et al, 1999).

Nevertheless, it is fair to say that most LEIT is promoted through strategies that are more farmer-centred than traditional extension activities.

Loevinsohn et al (2002) distinguish three types of learning that are relevant to promoting natural resource management skills. The first is 'reproductive' learning and corresponds to the rules or recommendations that are associated with much conventional extension. The second type of learning is 'constructivist', and is found in situations where farmers need to develop and adapt their own solutions. The third type is 'transformative', and includes the inductive development of new principles that help farmers to deal more effectively with future resource management challenges. A large part of the promotion of LEIT is based on constructivist and transformative learning skills. There is also a common presumption in many of the programmes for LEIT that some of the participating farmers will assume formal or informal responsibilities for further diffusing the knowledge that has been generated, at times assuming the role of farmer extensionist.

Although the development and promotion of a wide range of agricultural technology can take advantage of such methods, it is fair to say that LEIT projects have been particularly innovative in these areas. There are a number of formal extension or development methods that are strongly associated with LEIT development. Most LEIT activities are organized around group methods, and there is a wide range of examples. Many LEIT projects include opportunities for social learning; the FFS methodology is an outstanding example. In addition, there have been advances in developing farmer participation in problem diagnosis and technology development, and several formal methodologies exist. An additional characteristic of many LEIT efforts is an aspiration that, once useful principles or techniques are identified and introduced on a small scale, local farmers can take increasing responsibility for their diffusion, and LEIT projects often make provision for this strategy.

Group methods

Group methods are, of course, not the exclusive preserve of LEIT activities; as Pretty (1995) points out, many Asian countries promoted uniform Green Revolution packages through local groups (often formed or mandated for that purpose). But the group approach is particularly relevant to low external-input strategies for a number of reasons (Pretty, 1995). First, resource management often requires more than the efforts of individual farmers; for instance, communities need to agree on common strategies for challenges such as controlling soil erosion, and some pest management techniques require coordination among neighbouring farmers. The effective management of common property resources can play an important role in environmentally sound development strategies. In addition, low external-input strategies often go beyond the goal of introducing particular technologies and seek to build, or rebuild, local institutions. The recent interest in the theme of social capital means that group formation in agricultural development has assumed even greater importance (see Chapter 3). Finally, to the extent that LEIT requires the application of principles and the local adaptation of technology, it usually makes sense to organize group activities, not only because they are more efficient but also because they promote the interchange of ideas and experiences.

An example of group activity for improving common property management is the Landcare Movement, which originated in Australia and represents a productive collaboration between farmers and conservationists (Roberts and Coutts, 1997). Landcare groups comprise local farmers and other members of the community interested in land management. They meet to discuss common problems and engage in a range of activities, including teaching and training, trials, and interventions in areas such as soil conservation and wetlands management. Many of these activities are eligible for state funding and the Landcare groups also participate in the development of plans for integrated catchment management under the aegis of the state government. The innovation has been transferred to a region of the Philippines, where a Landcare Association and local groups have been formed (Mercado et al, 2001). The local groups include farmers, community leaders and extension agents, and they are able to take advantage of a decentralized government funding initiative related to environmental protection to support their activities.

Social learning

Social learning is a term applied to the transmission of information in a social context; in agricultural extension, it often refers to the potential of participants in a group activity to learn from each other. One of the best known innovations for introducing LEIT is the farmer field school (FFS), an excellent example of an opportunity for social learning. The development of an effective IPM strategy for rice in Asia is based on work in Indonesia that led to the emergence of the FFS concept. In order to appreciate the rationale behind lowering pesticide use and allowing ecological processes to re-establish their regulation of pest activity, farmers needed to learn more about pest–plant and pest–predator interactions; the FFS meets this challenge (Kenmore, 1996). It fosters discovery learning by facilitating farmer opportunities to observe and discuss important ecological relationships in the field. A FFS typically includes about 20 farmers and a trained facilitator and meets once a week during the cropping season in a designated field. Farmers spend part of each session observing pest and predator behaviour, drawing diagrams of the relationships they uncover and debating their implications. They are encouraged to make 'insect zoos' (collecting certain pests to observe their behaviour or life cycles) and to conduct simple experiments (such as mechanically cutting a certain proportion of leaves early in the crop cycle to mimic early insect damage and to observe the recuperative capacity of the rice plant). Each FFS also includes exercises in group dynamics to promote a group spirit and foster collaboration.

During the 1990s more than 75,000 FFS were conducted in Asia (Pontius et al, 2002). The FFS strategy is now being applied in a wide range of settings for purposes well beyond the extension of IPM. The United Nations Food and Agriculture Organization (FAO) is promoting the concept of integrated production and pest management (IPPM) through FFS in a number of African countries, where farmers are exposed to a range of principles in crop production (Okoth et al, 2003). There are worldwide examples of the FFS approach being applied to programmes in livestock health and production, soil fertility, community forestry, gender, HIV/AIDs prevention and other topics (CIP-UPWARD, 2003; *LEISA Magazine*, 2003).

Farmer participation

Farmer participatory research has been a part of many formal agricultural technology development strategies, at least since the emergence of farming systems research (FSR) during the 1970s. There are many variations and they encompass subjects well beyond LEIT. Biggs (1989) prepared a classification of types of farmer participation in public FSR programmes, and Robert Chambers and others helped to draw attention to wider possibilities for farmer involvement in technology generation (Chambers et al, 1989). A particularly useful guide for helping to organize practical farmer experimentation in the context of community-based projects was developed from the experience of introducing resource-conserving technology in Central America (Bunch, 1982). Braun et al (2000) compare the development of local agricultural research committees (a technique used in several Latin American countries for developing community-level experimentation capacity) with the experiences of FFS and find that each method has particular strengths and advantages.

An exceptionally comprehensive set of guidelines for developing options for integrated soil fertility management is found in the participatory learning and action research (PLAR) methodology. The objectives are to develop useful innovations in soil fertility management based on local resources and knowledge and, where appropriate, external inputs. PLAR was developed through a series of field experiences in Africa (Defoer and Budelman, 2000). The process, based on a series of activities with farmers over the course of several growing seasons, is an example of joint learning and experimentation, involving an interdisciplinary team of facilitators and farmers, and demanding a considerable amount of time from all participants. After focus communities have been selected, the first phase of PLAR is a set of diagnostic exercises, including analysis of the local land-use system, communication networks and soil fertility management strategies, including a classification of farm types and drawing resource flow maps for a small representative sample of farms. These farmers play a key role in the subsequent phases of planning and experimentation, and are selected to be representative of important farm types as well as because they are interested in experimentation and have good links with other farmers. They constitute a formal committee, with elected officers. The planning phase includes a farmers' workshop, exchange visits with villages known to be using innovative techniques, and the development and presentation of plans for experiments to be managed by participating farmers in the coming season. During the experimentation phase, the facilitators assist farmers to design simple experiments, record data and discuss emerging results with other participants. The final evaluation phase includes farmer meetings to discuss the season's experimental results and to plan a schedule for the subsequent season.

Farmer-led extension

Intensive methodologies such as FFS or PLAR, which direct considerable facil-itation resources towards a small number of communities, are often defended on the basis that once extension services have gained experience with the concepts in an initial period, they can adjust the methodologies to local resources. A complementary expectation is that farmers who have been through this process can serve as catalysts to help neighbouring villages initi-ate their own work. Extension services can make sure that experienced farmers have the resources to offer advice to others. A number of national FFS programmes in Asia have devoted resources to training and supporting farmer facilitators, who can take the place of extension agents in leading FFS and thus extend the methodology at a lower cost; however, the success of these efforts has been the subject of recent critical analysis (Quizon et al, 2001). Beyond these large-scale efforts, many other initiatives in promoting low external-input agriculture include farmer-led extension or farmer promoters in their strategies (Bunch, 1982; Scarborough et al, 1997; Holt-Giménez, 2001). In some cases, projects pay a small stipend to local farmers who have been identified and trained to play a formal extension role. In other cases, the projects simply try to increase contact between participant farmers and their neighbours in the hope that this will facilitate the diffusion of the innovations. In many cases, there appears to be an assumption that the experience of working in a group and generating LEIT will serve as a spark that motivates farmers to share their technological and organizational experience with others.

Summary

The purpose of this chapter has been to illustrate some major examples of LEIT, including many of those that will be discussed in Chapters 4 to 7. It is hoped that even this brief presentation has been sufficient to confirm that LEIT includes an exceptionally wide range of technologies, many characterized by multiple agronomic functions, complex economic and management attributes, and vari-able potential for addressing farmers' priorities. Many examples of LEIT offer significant advantages for farmers; but technologies need to be matched to par-ticular circumstances and adapted to local conditions and management. Without sufficient technical information and testing in specific contexts, exam-ples of LEIT may be ineffective, inappropriate or counterproductive.

The benefits of any technology must be balanced against the costs. LEIT is usually characterized as labour-intensive; but this review has shown such a clas-sification to be oversimplified. Some heavily promoted examples of LEIT have ultimately been rejected by farmers because they require too much labour. However, other types of LEIT may require only a significant initial labour investment, while labour demands in subsequent seasons are much lower. Other examples of LEIT require little if any additional labour and some are labour-saving. The timing and organization of labour requirements are also important considerations, further complicating decisions about LEIT.

The other characterization frequently associated with LEIT is its information requirements. Farmers' decisions regarding almost any type of technology usually involve the consideration of trade-offs; but LEIT's complexity makes it particularly subject to a range of difficult management decisions. LEIT may require the sacrifice of some crop land or may contribute to the recovery of previously unproductive fields; it may take resources previously used for one purpose and direct them elsewhere (for example, using crop residues for mulch rather than fuel); and it may provide an additional source of crop nutrients, but at the cost of some land. Farmers need technical information, access to the experience of others, and the opportunity to test and adapt new techniques. Even 'simple' technologies such as an improved fallow require locally adapted management skills, and more complex examples such as conservation tillage or SRI require several seasons of experience to master.

Such learning demands an investment; the intensity of many extension innovations for LEIT (such as FFS) illustrates these demands. Farmers must dedicate sufficient time to learning new principles and techniques (and are often expected to dedicate further time to sharing this information with their neighbours). The availability of that time depends, in turn, upon labour patterns and demands, and the economic returns from alternative uses of labour. Thus, labour and information are still key factors for understanding the career of LEIT. The following chapter examines the relationship of labour and information to agricultural technology development.

Labour, Information and Agricultural Technology

Robert Tripp

Introduction

Low external-input technology (LEIT) is often described as both labour- and information-intensive. As the review of these technologies in Chapter 2 has shown, such characterizations do not necessarily represent the exceptionally diverse nature of LEIT, the information and labour requirements of which vary according to technology, farming system and farmer experience. Nevertheless, it is broadly accurate to view LEIT as an attempt to lower dependence upon external inputs by placing greater reliance upon the utilization of locally available resources, and this usually has some implications for the deployment of labour and the management of information.

The purpose of this chapter is to provide a brief review of the literature related to labour and information and their roles in the diffusion of new agricultural technology. This discussion sets the scene for Chapter 4, which reviews the literature specifically related to the uptake of LEIT.

The chapter examines a range of sources; but there are two major contributors to the review. One is the extensive literature related to the adoption of technology (mostly generated by agricultural economists) and the other is the relatively recent interest in sustainable rural livelihoods, particularly those studies that focus on human and social capital and rural livelihood diversity.

Studies of the adoption of conventional technology offer many relevant points of comparison for our interest in the utilization of LEIT. The adoption literature provides insights on labour requirements, farm access to labour resources, sources of information and the role of education in technology diffusion. It is often possible to correlate labour and information assets with the patterns of uptake of a technology. In addition, formal adoption studies often examine economic factors such as farm size or wealth, land tenure or access to credit (e.g. Feder et al, 1985) that are important not only for understanding the rationale of technology adoption, but also for assessing the equity of access to the technology. The almost inevitable linkage between certain resources or skills and preferences for new technology means that most innovations will be 'biased' in one way or another. In his comprehensive review of the diffusion of all types of innovations, Rogers (1995) concludes that the common (although not inevitable) consequence of technology diffusion is to favour the better off, thus widening socio-economic gaps. There is considerable hope that at least some types of LEIT can be directed towards those farmers for whom much conventional technology is not appropriate.

Although the formal adoption literature provides a number of insights to help understand the diffusion of technology, it also has several weaknesses for the purposes of this study. In the first place, it is impossible to make neat divisions between the 'economic' and the 'social'. Farmers' decisions about technology may ultimately be based on an assessment of costs and benefits; but the data that enter such calculations are socially mediated. The opportunity cost of labour is determined by individual household circumstances and aspirations. An understanding of technology performance depends upon information from others, including neighbours and public servants. Willingness to invest depends upon perceptions of income-earning strategies and an assessment of the contribution of farming to the household's livelihood. Such socially mediated factors are often difficult to assess in formal studies, and our knowledge of their importance remains incomplete. As Feder et al (1985) point out, the mathematical functions appropriate for formal adoption studies may not correspond to any real underlying decision behaviour.

In addition, adoption studies may not always provide practical advice. As Lockeretz (1990) warns, it is easy to fall into the trap of mechanical analysis where statistical significance may be confused with importance, causality is not thoroughly explored, and the emphasis may be on details while larger questions remain unexamined. Although the methodology of a formal adoption study provides opportunities for an objective assessment of certain aspects of the career of a new technology, this must be complemented by other types of inquiry and analysis.

The recent literature on sustainable rural livelihoods provides further insights for understanding the relationship between social factors and the diffusion of agricultural technology, emphasizing, in particular, the role of human and social capital. These terms provide an attractive shorthand for the capacities and resources that are key to understanding the performance of agricultural technology, and have the additional advantage of widespread usage, in part because of the popularity of the sustainable livelihoods (SL) paradigm. Pretty et al (2003) have produced a framework in which progress towards sustainable agriculture is assessed by the protection and accumulation of the five assets (natural, social, human, physical and financial capital) that feature in the SL framework. Pretty and Ward (2001, p212) ask: 'to what extent ... are social and human capital prerequisites for long-term improvements in natural capital?', which is a useful way of framing the basic questions that motivate this book. Access to the terminology of the SL framework also draws attention to livelihood diversity and its relation to the utilization of household labour, which is one of the key factors related to the uptake of LEIT. For instance, Bebbington (1999) has described how the SL framework can be used in thinking more comprehensively about strengthening peasant livelihoods and the deployment of household assets for both farm and non-farm activities.

The SL framework can contribute to a broader examination of technology utilization than a simple adoption study. But despite the useful contributions of the SL framework, there are also questions about the practicality of much of the SL literature and the lack of precision in concepts such as social capital and livelihood diversity. It is necessary to specify key manifestations of these concepts as precisely as possible. Thus, neither the results of formal, focused adoption studies nor the conclusions of broad-based SL analysis are sufficient, on their own, to provide an adequate understanding of the utilization and impact of new technology. The review must draw from as wide a range of sources as possible.

This chapter begins with a discussion of the labour component of agricultural technology, drawing on a range of literature that shows the role that labour supply and allocation play in farmers' decision-making. The discussion also includes an examination of the interplay of agricultural labour with broader decisions on household labour allocation. It then examines the general role of information in agricultural technology, as well as specific evidence from adoption studies. Information flow and management depend upon farmer capacities, manifested in both human and social capital. The discussion then turns to examine human capital, including evidence from conventional adoption studies (usually focused on formal education), as well as other aspects that are more difficult to quantify, such as innovative capacity and indigenous technical knowledge. The discussion finishes with an examination of social capital, beginning with an attempt to draw some useful boundaries around the term and then proceeding to examine evidence related to agricultural innovation.

Labour

Farm labour

In his thorough description of tropical farming systems, Ruthenberg (1976, p26) points out that it is:

> ... those processes that offer higher returns on labour that many farmers actively seek, and the development of small-scale production must largely depend on the discovery and adoption of practices that will substantially improve returns to work. From the farmer's point of view, labour productivity and work rationalization are considerations of great importance, not only because labour is a major input but because its use is the subject of acute personal experience.

His detailed tables showing inputs and outputs of various types of traditional production systems (including shifting cultivation, fallow systems and permanent cultivation) illustrate an exceptionally wide range of labour input per hectare (and output per hour). The variation can be explained in part by the length of the growing season, the tasks performed, participation in cash crop markets and the availability of labour-saving technology.

It is important at least to note at this point that farm labour is not only a candidate for technical and economic analysis. Despite being a 'subject of acute personal experience', it can also be a source of pride and an exercise of carefully honed skills. Van der Ploeg (1990, p27) describes 'the craftsmanship of farm labour ... the continual observation, interpretation and evaluation of one's own labour in order to be able to readapt it. This process is in marked contrast to industrial labour, where the labour process can usually be broken down, quantified, predicted, and therefore planned and controlled.' Such craftsmanship in farm labour is recognized within rural communities, serves to distinguish among households, and contributes to establishing reputations and positions in rural society. The largely economic treatment of labour in what follows is balanced by a review of the human capacities and social organization of farm labour later in the chapter.

Labour is a primary input to virtually all agricultural production. The availability and cost of labour help to determine the type of technology used in farming. Hayami and Ruttan (1985) have shown how the relative availability of land and labour 'induce' technological development. The type of technology developed by farmers, public research and industry tends to correspond to relative scarcities of labour and land. Technological change in those countries or areas with relatively scarce labour tends to favour innovations that increase returns to labour (such as mechanization), while relatively labour-abundant farming areas follow a land-saving pathway (such as higher investment in soil fertility).

The importance of reducing labour costs does not, of course, preclude the possibility of increasing total labour investment, when conditions are appropriate and the returns are attractive. The path of agricultural development has often been characterized by increasing investments in labour; the increasing complexity of Asian rice production is an example (Bray, 1986). For instance, in Tokugawa, Japan, the use of commercial fertilizers, improved threshers, multiple cropping and crop diversification led to a significant increase in the use of farm labour. The increased labour responded to the requirements of new technology (including external inputs) and market opportunities. Similarly, agricultural change in the Cauca Valley of Colombia during the 1970s, spurred by improved transport systems and dynamic agricultural markets, saw a significant increase in labour invested per hectare on farms growing a wide range of crops, including coffee, vegetables and traditional staples. Average farm size was growing smaller; but increasing investments in labour and inputs made the farms more productive (Reinhardt, 1983).

The Green Revolution in Asia is a particularly important case of technological change leading to increased use of labour, much of it hired. The development of short-statured rice and wheat varieties, in conjunction with increased use of fertilizer and irrigation, made possible significant increases in yield that reversed the trend towards increasing food grain imports. In many cases the new varieties allowed for multiple cropping (e.g. growing two crops in a year where only one had been possible before), generally required more (and more carefully timed) weeding and other crop management, and led to increased demand for harvesting and threshing labour in response to the higher yields. A review of the adoption of modern rice varieties in a number of Asian countries found, with one exception, a significant increase in labour use per hectare (David and Otsuka, 1994).

Farmers are thus willing to invest in additional labour when the incentives are appropriate. The extension of seed-fertilizer technology to African farmers by Sasakawa-Global 2000 is an example. In Ethiopia, those maize farmers who adopted the high-input package invested 80 per cent more labour per hectare than their neighbours because of the increased demands of row planting, input application and harvesting the significantly higher yields (Seyoum et al, 1998). But the return to labour is a crucial factor influencing the adoption of any new technology, something that colonial and post-colonial agricultural development efforts in sub-Saharan Africa often overlooked (Collinson, 1983). For instance, there are many examples in which recommendations for farmers to increase the number of times a crop is weeded fall on deaf ears, not because the farmers are lazy but because they are more sensitive to assessing returns to labour than are the extension agents issuing the textbook advice. Similarly, recommendations for more efficient fertilizer management through techniques such as split applications or deep placement may not be accepted because of their extra labour implications.

In addition, the absolute amount of extra labour required by a new technology may be less important than the timing of that labour within the cropping season. The temporal allocation of labour, and particularly the mitigation of seasonal bottlenecks, is crucial to a productive farming system. The Kofyar of central Nigeria are very successful agriculturalists who grow a range of subsistence and commercial crops through an exceptional investment in farm labour (Stone et al, 1990). Part of their success depends upon the ability to marshal labour for peak periods of demand, such as early in the season when planting the first millet crop has to be combined with field preparation for other crops, or the harvest of the early millet at the same time as yams and sorghum have to be weeded. The labour investment is made by a combination of household effort, neighbourhood work groups and exchange labour. The farmers tailor lower-priority crop management tasks to fit relatively slack times.

Farmers in northern Côte d'Ivoire who grow cotton and food crops try to respond to labour bottlenecks through a variety of techniques for spreading out or reducing labour demands (Bassett, 1988). Like the Kofyar, they also use communal labour; but the increasingly market-oriented production patterns contribute to the erosion of these customs. Cultural norms (such as traditionally prescribed rest days) are also subject to reinterpretation. Cropping patterns may be modified, and those crops that demand more labour (such as yams and millet) begin to lose favour. Labour-saving inputs, such as herbicides, become more common and the amount of hired labour increases.

Communal labour has traditionally been an important way of overcoming labour bottlenecks, substituting the pleasures of social interaction for the drudgery of individual toil and fulfilling social obligations. Communal labour (work parties, exchange labour, etc.) continues to be important for many societies, although these customs are undoubtedly breaking down in the face of wage labour markets. It is also important to recognize that although communal labour is often an exceptionally effective way to address labour demands, it is not always a sign of egalitarian farming communities. Watson (2004) provides a contrast between two intensive agricultural systems in Eastern Africa that is particularly relevant for our interest in resource-conserving technology. In Kenya, the Marakwet use communal labour to construct and maintain a network of irrigation canals that contribute to a particularly productive agriculture. Because the irrigation water provided by this is available to all, the system is marked by little social stratification. In contrast, the Konso of Ethiopia construct drystone-walled terraces and practise other soil and water conservation techniques. Much of this is accomplished through various types of work group; but because it is applied to individually owned fields, those with greater wealth can attract more labour and further enhance the quality of their land. Watson's study showed that clan leaders attracted nine times more extra-household labour than their neighbours, and that most of this investment was devoted to conservation measures.

Hired labour is an increasingly important strategy for addressing farm labour demands. Any thought of evaluating a technology (including LEIT) based on the assumption that farming communities are composed of undifferentiated, self-sufficient peasant households is unrealistic. The use of hired labour in farming is widespread, and the consideration of the extra labour requirements of a new technology often relies upon the nature of the local labour market. In many cases, labour is as much an 'external' input as fertilizer, requiring access to cash. When Amanor (1994) asked a sample of small-scale food crop farmers in a village in Ghana what they would do if they were given a loan, 84 per cent of them said they would invest the entire amount in hired labour.

The majority of the extra labour demand created by Green Revolution technology in Asia was met by hired rather than family labour (Lipton with Longhurst, 1989). In the Indian Punjab, the Green Revolution was responsible for only modest increases in daily agricultural wages, but resulted in significant increases in number of days of employment per year. The demand often led to the establishment of yearly labour contracts for former casual labourers (Leaf, 1984). Agricultural wages are affected by a number of factors, including the productivity of the farming system, the efficiency of local agricultural markets, the development of non-farm labour opportunities and farmers' own decisions to pursue off-farm opportunities. An examination of changes in south India over a decade showed that total employment in rice farms dropped (largely due to mechanization), the use of family labour on small and large farms increased, real agricultural wages increased slightly but not consistently, and total household income for landless labourers rose significantly (mostly because of non-farm opportunities) (Hazell and Ramasamy, 1991).

As wage rates for rural labour rise because of either demand from the agricultural sector or competition from non-farm opportunities, farmers turn to labour-saving technology to cut costs. Studies in India showed that during the early years of the Green Revolution, a 1 per cent increase in technology-induced agricultural output led to an additional 0.4 per cent growth in agricultural employment. But as farmers sought labour-saving technology (at least partly in response to increasing rural wage rates), by 1990 the equivalent ratio had fallen to 0.1 per cent (Kerr and Kolavalli, 1999).

One manifestation of this trend is the widespread use of herbicides in many Asian rice farming systems (Naylor, 1994), where non-farm employment opportunities have increased rural wage rates, lower rice prices have made farmers more cost conscious, and the replacement of transplanting by the labour-saving practice of direct seeding (itself a reaction to higher labour costs) has called for different weed-control strategies. Labour-saving technology may have other advantages; mechanized land preparation and planting, for instance, can provide much more precise control over the timing of farm operations, leading to higher yields. In rural economies where agricultural markets are stagnant but other employment activities exist, the use of external inputs may, in effect, save household labour that has a higher opportunity cost. This explains the paradox described by Low (1986) in Swaziland, where households that only produce maize for subsistence nevertheless invest heavily in hybrid seed and fertilizer; earnings from migratory wage labour are invested in these technologies that support the labour-deficient farm.

Off-farm labour

The availability and costs of labour for agriculture depend to some extent upon the nature of non-farm employment opportunities available to rural households. The labour requirements of any new technology may be interpreted differently by various types of farmers, depending not only upon their access to household labour (or ability to hire labour), but also upon their alternative employment opportunities. These alternatives can draw them away from the farm, but may also provide extra cash resources to hire local labour. Non-farm income is not a new phenomenon, but its importance is increasing even where it has traditionally been a major livelihood strategy, as Blaikie et al (2002) discuss with regard to western Nepal. Between the 1970s and the 1990s, non-farm sources became a more important component of the incomes of labouring households in this region of Nepal. For peasant households, the proportion with some type of non-farm income remained unchanged (about two-thirds); but wealthier peasant households expanded their reliance upon non-farm income, resulting in a positive relationship between agricultural and non-agricultural incomes.

The deployment of farm household labour is receiving increased attention due to growing interest in rural livelihood diversity (Ellis, 1998; Bryceson et al, 2000). There is growing recognition of the importance of non-agricultural income to rural households; a recent study of villages in Morogoro, Tanzania, found that approximately half of household income came from sources other than agricultural production (Ellis and Mdoe, 2003). The availability of non-agricultural income opportunities may reduce the amount of labour available on the farm. On the other hand, where agriculture is not profitable enough to offer a sufficient wage but there are relatively few outside alternatives, farm labour shortages and rural unemployment may, paradoxically, co-exist.

Although family traditions and reputations in rural communities undoubtedly depend upon the 'craftsmanship' of farmers' labour on their own holdings (van der Ploeg, 1990), rural economic realities challenge the assumptions of earlier agrarian sociology or the current oversimplified images of the peasant farmer promoted by some non-governmental organizations (NGOs). Van der Ploeg himself provides a striking illustration of these realities in a description of a highland Peruvian farming community, where only 9 per cent of the farmers work only on their own farms. The poorest group of farmers combine subsistence production with wage labour in order to survive. Other farmers invest part of their time in wholesale trading and look to this as their principal economic strategy, seeking to reduce labour and other investments in the farm. Another group is involved in enterprises such as petty trade, which offer less scope for expansion, and they channel much of their earnings to further intensifying agricultural production.

Labour migration can take different forms. In Mexico, Grindle (1988) describes several contrasting environments. In a village in Michoacan, households have, for historical reasons, sought opportunities off-farm, and many young men go to 'the other side' (the US). The local economy has become dependent upon income from migratory labour, and although the modest agricultural success of some nearby communities indicates that investment in farming may have pay-offs, the incentives are simply no longer there. In contrast, an area in the Bajío region exhibits a flourishing commercial agriculture; but it comes at the price of considerable differentiation in wealth and access to productive resources. In particular, the availability of irrigation to some farmers has allowed them to expand and become more productive, while other households are sources of farm labour on the irrigated properties during the growing season and exporters of short-term migratory labour to other areas of Mexico during the slack season.

Although it may still be possible to combine own-farm labour with off-farm migration after the cropping season, the concentration of rural resources is making this less of a possibility. Many farms may simply not be viable, and an absence of external opportunities means that farmers end up seeking day labour with wealthier neighbours, as in the *ganyu* labour system in Malawi (Whiteside, 2000). Such systems can contribute to the downward spiral of small farms since opportunities for labourers occur at precisely the times when they should be working on their own farms. The unequal availability of off-farm income for households with varying resources may contribute to increasing land concentration. Murton (1999) shows a recent increase in the inequality of landholdings and agricultural incomes in Machakos, Kenya, fuelled by off-farm income; the 20 per cent of households earning the highest non-farm incomes control 55 per cent of the land.

The degree to which off-farm income is a substitute for, or addition to, farm investment depends upon several factors. Ellis (1998) reviews a number of studies on the subject; many show that diversification to non-farm activities is often the result of a lack of agricultural opportunities, and that such diversification further weakens incentives for farm development. There are cases, however, where off-farm income can support farm development. As Reardon et al (2000) discuss, this may partly depend upon whether significant off-farm opportunities require education and the nature of access to that education. One study reviewed (from Kenya during the 1970s) shows how relatively equitable access to education combined with growing off-farm opportunities fed productivity-enhancing and scale-neutral agricultural investment. On the other hand, there are many examples where those households with better qualified members can take better advantage of off-farm opportunities, often contributing to a growing disparity in agricultural success, as well.

The impact of off-farm income on farm livelihoods and strategies is a complex issue. Reardon et al (2000) summarize a large number of case studies that show quite variable patterns. In some areas of high agricultural productivity (such as irrigated Punjab), higher-income households obtain a lower share (and even, at times, a lower absolute amount) of income from off-farm sources. Agriculture remains an important contributor to advancement. In many of the other cases, however, (including the majority for Africa), both relative shares and absolute amounts of income from off-farm sources rise for the wealthier households; not surprisingly, aspirations in these cases tend to focus on non-farm opportunities.

Even when off-farm employment contributes cash for farming activities, there is a question of the skills of those who remain on the farm and their capacities to take up new practices. If migrants tend to be better educated and male, does the older, and increasingly female, farm labour force have the time or the skills to invest in new technology? The issue of the feminization of farm labour, as men increasingly seek off-farm employment, is important for many parts of Africa and Asia (Ellis, 1998; Harriss-White, 2003).

A broader view of labour

The skills and incentives of those managing the farm are a key issue for the consideration of LEIT. The labour requirements of a technology may be of several types. Some technologies may require a high initial labour investment and then modest maintenance in subsequent seasons, and these may even offer a labour saving over the long run. Other technologies require labour investments each season. The costs (and opportunity costs) of that labour, and the expected returns, play a large role in determining the farmer's interest in the innovation.

In addition, part of the decision regarding a new technology is related to the learning costs associated with gaining acquaintance with the techniques or principles, or making periodic adjustments and adaptations in practices. This is not usually counted as labour; but there may be a significant investment needed to obtain knowledge about a technology, with implications for both the time requirements for adoption and the quality of the labour demanded. Such learning costs are not easy to quantify, but can be important determinants of adoption. Learning costs are also important when the technology is related to more radical changes in cropping patterns or farm management, or when the benefits are realized only in the long term or emphasize resource conservation rather than immediate economic gain.

Thus, while choices about innovation are related to labour requirements, they often go well beyond a simple comparison of hours invested with expected returns. The examples are not limited to LEIT or to developing economies. Perceptions of the farm and its future, the nature of agricultural work and alternative livelihood options all play a role in the decision. The results may be difficult to understand simply in terms of conventional economic analysis. When Kentucky farmers first saw the results of cost-saving and resource-conserving reduced tillage systems, they were unimpressed, preferring the aesthetics of the neatly tilled rows that they were used to (Choi and Coughenour, 1979). Smith (2002) points out that smaller farms in the US derive a high proportion of their household income from non-farm activities, and this division of attention may make them less likely to take an interest in resource-conserving, but management-intensive, techniques, such as precision farming or integrated pest management (IPM). Attempts to explain the wide-spread (and enthusiastic) uptake of genetically modified herbicide-tolerant soybeans in the US have found little impact upon net farm returns, and have had to fall back on farmers' explanations regarding the 'simplicity and flexibility' of the technology (Fernandez-Cornejo and McBride, 2002). Further analysis has shown that those who adopted herbicide-tolerant soybeans tended to be smaller farmers and that adoption was significantly related to increased levels of off-farm income (Fernandez-Cornejo et al, forthcoming).

In summary, the decisions that farmers make about LEIT, or any other type of technology, are based, in part, on strategies regarding labour deployment. These strategies depend not only upon the amount and timing of labour required for a particular innovation, but also upon an assessment of the skills available for farm management, perceptions about changes in routines of work or farming goals, a particular vision of farm stewardship, and an evaluation of farm and non-farm livelihood opportunities. And these perceptions are, in turn, conditioned by the information available to the farmer about new technology and its management. That information is mediated by factors related to human and social capital; but before introducing those topics it will be useful to review the characteristics of the information used in agricultural decision-making.

Information

The terms information and knowledge are often used interchangeably. Although this book does not attempt to make a precise distinction between the two, it is useful to point out some differences in the usage of the terms. Information is often treated as a commodity that can be transferred or shared, while knowledge implies a human effort to digest and gain control of information (Brown and Duguid, 2000). This section reviews studies on the availability and transmission of agricultural information, and following sections discuss how human and social capital transform that information into knowledge. Information may be something as straightforward as the identity of a fertilizer dealer or as complex as the way in which insect population dynamics can be used to control a pest, using its natural enemies. Information is something that is discovered by particular actors (a farmer in a field, a scientist in a laboratory) and is transmitted from one actor to another. We are particularly interested in the efficiency of that transmission process. Knowledge, on the other hand, is the ability to take advantage of that information and it resides with individual actors (farmers, extension agents, plant breeders, etc.).

As already noted, LEIT is often described as information-intensive technology. There is no doubt that many of these techniques cannot be utilized in a mechanical fashion and require careful, experienced management. Whether LEIT requires more knowledge than other types of agricultural practices, or is uniformly demanding of information resources, is more open to question. Nevertheless, the successful diffusion of LEIT requires that farmers have access to certain types of information, and much of the rest of this chapter is devoted to exploring how human and social capital contribute to the way in which information is exchanged and utilized by farmers.

As S. Jones (2002) points out, an understanding of the decisions that farmers take regarding resource conservation involves examining several distinct issues, including farmers' perceptions of production problems, their knowledge of techniques, their incentives for investment and their access to resources to make the investment. There is, thus, a diversity of types of information relevant to LEIT, including awareness of particular problems and practices, the principles behind innovations, where to turn for advice, and an appreciation of long-term environmental relationships. These various types of information must be accessible if LEIT is to have a chance of taking hold. We are interested in how farmers acquire this information.

Information has an important role in all types of agricultural systems, and it is not clear if LEIT is more information-intensive than other production practices. There is evidence that information has an increasingly central part to play in agricultural development, in general. Robert Shapiro, the former Monsanto chief executive, once offered a rather blunt description of biotechnology as representing a trend in agricultural history in which '"information" was replacing "stuff"' (cited in Charles, 2001, p270). Byerlee (1998) argues that farming in industrialized countries and in Green Revolution areas in the South will have to pass from the stage where knowledge is 'embedded' in the inputs provided to farmers to a stage where information itself becomes a more important input. The provision of information will have to shift from general area-based recommendations (for example, for fertilizer use) towards qualitative season- and site-specific advice on the types of inputs (external or locally produced) that might be applied, and eventually to quantitative assessment based on criteria such as soil tests or pest counts. His analysis emphasizes the organization and costs of providing that information (from research, extension or the private sector), but acknowledges that farmers' education is a key parameter in lowering the costs of information delivery (and a potential discriminating factor in determining the equity of that information provision).

It is common to hear that LEIT substitutes information for chemicals, so presumably education, skill or experience may play an important role in facilitating access to this type of technology. However, these relationships have not been carefully explored. Lockeretz (1991, p97) reviewed the literature and concluded that 'discussions of the role of information in reduced-chemical production barely go beyond the assertion that it is important'. He points out that we need to know whether such technologies are information-intensive solely to develop or also to adopt; if the latter, whether the additional information needs to be acquired only once or whether it needs to be continually sought and updated; and whether the acquisition of the additional information requires only an investment of time or also implies new skills.

Any discussion of information provision naturally includes the organization of external information sources, which has traditionally meant extension services. The effectiveness of extension, particularly in reaching resource-poor farmers, has been the subject of considerable controversy. One of the most prominent debates in recent years has focused on the Training and Visit (T&V) system of extension, promoted by the World Bank on the basis of initial praise (Bindlish and Evenson, 1997) and then abandoned after increasing doubts about its costs and effectiveness (Gautam, 2000). T&V brought considerable order to extension management, but was based on a very mechanical and top-down model of information provision. This is not the place to review conventional extension strategies, especially as LEIT is associated with a number of alternatives for generating and diffusing information that rely more upon farmer participation and initiative. Thus, it is more appropriate to concentrate on farmers' motivations and abilities for acquiring and sharing information. Further justification for this approach comes from a review of studies of the effectiveness of conventional extension strategies, which found evidence of impact in a number of cases, but pointed to two serious problems in attributing that impact. The first is differences in farmers' interest and abilities in seeking out information. The second is the fact that once externally provided information becomes available, it is often transmitted more effectively among farmers (which explains why farmers, in the majority of studies reviewed, cited other farmers, rather than extension, as their major source of information) (Birkhaeuser et al, 1991).

The literature on agricultural technology adoption provides some guidance on the nature of information flow among farmers. Feder and Slade (1984) point out that information-gathering may be 'active' (entailing costs for seeking experience or instruction) or 'passive' (taking place in an autonomous manner, often by observing neighbouring farmers). Several studies have attempted to determine the degree to which farmers can rely upon their neighbours' experience, as opposed to learning about a technology for themselves (e.g. Besley and Case, 1993). Heinrich (2001), examining adoption models in general, argues that the observed patterns are most consistent with 'cultural transmission' – that is, farmers copying their neighbours' behaviour rather than doing their own experiments. Fischer et al (1996) examine adoption models where information becomes more available as more farmers try a new technology. They apply the model to the adoption of new wheat varieties in Australia and conclude that, even if experience is overwhelmingly positive in one year, farmers need several seasons to become convinced of the value of the new variety. Cameron (1999) argues that because most adoption studies rely upon cross-sectional data, they overlook the importance of farmers' own experience over several seasons in shaping adoption decisions. Foster and Rosenzweig (1995) test whether neighbour experience can substitute for own experience, using Green Revolution data from India. They find that own experience (based on panel data for the farmer's previous year's planting) and neighbour experience (based on data for the practices of others in the village) are both important.

Thus, there is evidence that farmers use their own experience, experimentation and observation, as well as what they learn from interactions with their peers and other information providers. These two streams roughly correspond to the distinction between human capital (farmers' endogenous skills and capacities) and social capital (their links with others that provide useful information and support). These are the subjects of the following sections.

Human capital

Education

There are various approaches to the subject of human capital. Some are very broad; Sen (1999), for instance, sees human capital as a subset of his more inclusive concept of human capability – people's ability to lead the lives that they value and to take advantage of an expanded range of choices. Most studies of human capital include health and nutrition, but the discussion here will be limited to a person's stock of skills and productive knowledge. These are a function of both formal schooling or training and experience or knowledge gained during work or employment. Schultz (1964) was responsible for pointing out the key role of human capital in agriculture, and there is much evidence of a strong contribution of human capital to agricultural progress. Nevertheless, the relationships are not necessarily straightforward, and there are a number of uncertainties that are relevant to this review.

Formal education is practically an obligatory variable in studies of the adoption of agricultural technology. Feder et al (1985) review some of the evidence behind the often positive relationship between educational level and technology adoption or farm productivity. They find that farmers with higher education tend to be early adopters of new technology, but show that education may have little influence on productivity in areas practising 'traditional' agriculture, where new techniques play a minor role. Birkhaeuser et al (1991) point out that extension may be a substitute for education and that the impact of formal extension may be greater where education levels are low.

Although many adoption studies reveal a positive relationship between a farmer's education level and the uptake of conventional technology, the precise reasons for the relationship are rarely explored, and the fact that differences in this effect often vary according to the type of technology remains unexplained. What is it about education that makes it important for at least some technology adoption, and do these factors hold for the adoption of LEIT? For instance, in Ghana, Morris et al (1999) find that the adoption of modern maize varieties and row planting is significantly related to the farmer's educational level; but the adoption of fertilizer shows no relation to education. Why should this difference by type of technology exist (and although the approximately one year of education advantage of the adopters is statistically significant, is it of any practical importance)? Feder and Slade (1984) examine Indian farmers' knowledge and adoption of several technologies and find in each case a relationship between adoption and years of schooling, but in only one case (adoption of herbicide) a relationship with the farmer's score on a simple test of numeracy.

When formal education does have an influence on technology adoption, is it because of specific skills (such as numeracy), a change in worldview, better contacts with opinion leaders or other factors? Obtaining an answer is difficult, and it will certainly vary by situation; but the search is important because it helps to determine the degree to which investment in formal education might contribute to technological change on the farm. The fact that the best opportunities for off-farm income often require higher educational qualifications, and increased access to education may shift households away from farming, makes the issue even more complex (Reardon et al, 2000).

Innovation, experience and indigenous technical knowledge (ITK)

There are many other components of human capital besides formal education that affect farmers' access to technology, but these are rarely investigated. For instance, the capacity to take advantage of an innovation may be related to experience in farming. When this is examined in adoption studies, a common proxy is age. When such a correlation is significant, it may be positive (in cases where elements of experience or confidence make farmers more likely to test innovations) or negative (often related to conservatism or changes in economic strategies at a later stage of the household development cycle). But it is unusual to find any attempts to apply such evidence to improvements in information dissemination.

It is likely that the type of information required for LEIT will be more on the side of principles (to be adapted and applied) than on recipes (to be copied). In addition, detailed knowledge about one's own fields may be as important as knowledge of new principles when utilizing low-input techniques. This is especially the case for matters such as soil management techniques, which must necessarily be tailored to individual circumstances. In these instances, longer farm experience may contribute to the likelihood of making the best use of such principles. These challenges of adaptation are not confined to LEIT, and form the basis of practices in precision farming in agricultural settings where the precise level of external inputs is differentiated within and between fields by location- and plot-specific information (Cassman, 1999). In this case, however, externally provided techniques for information management (such as computer software and remote sensing) may partially substitute for farmer experience.

Another element of human capital, rarely examined in formal studies, is a farmer's capacity to innovate. An important issue here is deciding what exactly constitutes innovation. In Reij and Waters-Bayer's (2001) study of farmer innovation in Africa, innovators were described simply as those farmers who tried new things, leading to an inventory of everything from novel soil management techniques to unremarkable trials of new crops or varieties (the latter accounting for 44 per cent of the innovations recorded). Many of the most noteworthy innovations they describe are in the area of soil and water conservation (where individual adaptation is a key factor). But it is also important to understand the conditions and incentives for such innovation, and many of the new practices described in their study are in response to market opportunities or resource pressure, rather than being the products of idle curiosity. Innovators described in the review come from various backgrounds; but there is a tendency for them to be older, wealthier men, often with outside experience.

The concept of experimentation is more precise than innovation. In a study of farmer experimentation in Africa, Sumberg and Okali (1997) identified two conditions for an 'experiment': a farmer's initial observation of conditions or treatments and the observation or monitoring of subsequent results. They found a wide difference among individuals and sites in the propensity to engage in experimentation. They also distinguished between 'proactive' experimentation, in which there was a conscious effort by the farmer to create or control certain conditions for the purposes of observation, and 'reactive' experimentation, which had no systematically chosen objectives. Their study found that men are more likely than women to report experimentation, and that these men are likely to have more education and previous involvement with extension. It also found that experimentation was more frequent among people who saw themselves as full-time farmers, frequently (although not exclusively) engaged in commercial production.

If much of LEIT is based on traditional practices, perhaps less 'scientific' information is required. Success may depend more upon the acquisition or reactivation of indigenous technical knowledge (ITK), another important component of human capital. Murwira et al (2000) report how farmers in a rural development project in Zimbabwe were at first reluctant to discuss their use of traditional practices in activities such as pest control for fear of ridicule; it was only when the project confirmed that such knowledge could have significant value that farmers were willing to include the information in the project's activities.

Much has been written about ITK, providing examples of farmers' detailed understanding of the environment. The Hanunoo cultivators of the Philippines recognize many more distinctions in the plants growing in their region than do systematic botanists (Conklin, 1957). Traditional soil classification in Tlaxcala, Mexico, is a comprehensive yet flexible system that allows farmers to understand both fertility and soil management properties (Wilken, 1987). But the capacity to observe and classify does not always lead to practical knowledge. Although the Hanunoo recognize 20 main categories of insect pests in their swidden fields, they have few effective remedies in cases of severe pest attack. Similarly, Bentley (1989) found that Honduran farmers' ITK was best developed for describing plants and plant growth, and less adequate for understanding insect behaviour or the origins of plant disease. These lacunae may contribute to harmful practices, such as overdependence upon pesticides. Thus, ITK cannot be seen as a sure defence against environmental misman-agement. Wilken's conclusion usefully places ITK in context:

> It is ... condescending to view traditional farmers as sagacious husbandmen, imbued with infallible folk wisdom, in mystic harmony with the environment. In fact, they share with their industrialized counterparts the human propensity to respond to short-term opportunities while disregarding long-term costs, to misinterpret the durability and flexibility of agro-ecosystems, and generally to err (Wilken, 1987, p268).

Amanor (1991) describes the ITK of Krobo farmers in Ghana, particularly related to weed management and the dynamics of a fallow system that was challenged by increasing farming intensity. The contribution of the farmers' indigenous knowledge lies less in its capacity to inventory species than to confront a rapidly changing environment:

> The value of Krobo farmers' observations of the fallow does not arise from its efficacy as a traditional stable knowledge system. Its value lies in its attempts to wrestle with problems caused by the disruption of established practice, its lively and wistful reflection of processes at play in the environment, its observation and artic-ulation of problems it has not solved, and its interplay with an environment that calls for mastery (Amanor, 1991, p12).

The study contrasts the broad knowledge that farmers have of their environ-ment with the reductionist strategies of local development programmes promoting the introduction of particular agroforestry species for soil regener-ation. The farmers can certainly take advantage of scientific agricultural knowledge; but the programmes providing it need to recognize the broad experience and capacities of the farmers themselves.

The relationship between human and social capital

Human capital and social capital (the subject of the following section) are not perfectly distinct concepts. Information provided by formal education or the exercise of personal ingenuity are examples of human capital. But if the infor-mation is derived from links to other farmers, then we must also consider social capital. The potential of human capital may only be realized if adequate social capital is available. Distinctions between social and human capital may become blurred, especially if we shift between considering the type of information and the way in which it is transmitted. One farmer's formally acquired knowledge (an example of human capital) may be shared with other farmers through social networks (social capital), further strengthening the recipients' capacities (human capital). If the information comes from formal extension, it may be difficult to decide if human capital (e.g. literacy) or social capital (for instance, links with extension agents) is more important. In one sense, human capital considers an end-product (human capacity), while social capital focuses on social networks and organization as a means of transmitting information and building capacity.

Early studies on social capital (e.g. Coleman, 1988) emphasized its role in building human capital – for instance, in supporting access to formal education. Lin (2001) reviews several European studies on job choices and income, showing how human capital and social capital are complementary: either formal educational qualifications or social links may be used in setting career paths. As Grootaert et al (2002, p26) point out in Burkina Faso, in cases of 'low levels of education ... social capital may act as a substitute for human capital and most of the knowledge needed to generate income may need to be acquired from social networks and associations'. We now turn to examine briefly some aspects of the concept of social capital.

Social capital

Definitions and uncertainties

Although social capital is a theme that has emerged only recently, it has captured the imaginations of many development specialists and has already found a place in the literature on LEIT. It is useful to review the concept in relation to agricultural change. In the first place, many observers point out that LEIT benefits from strong social relations within a community that help farmers to organize and learn new production techniques, as well as considering common strategies for activities such as soil recuperation or pest management. To the extent that LEIT requires a reorganization of farm labour patterns, social capital may be useful in organizing and directing that labour. A strong sense of community is also important for gaining appreciation of the environmental protection and enhancement objectives of LEIT. In addition, many who promote LEIT hold that a farmer group's successful experience in managing and adapting more self-reliant technology is an impetus to further organized efforts for improving farming practices and promoting community development. This section begins by examining the concept of social capital and then reviews examples where it has been applied in relation to agricultural practices.

Part of the reason for the widespread use of the term social capital is its niche in the sustainable livelihoods approach. A focus on the various assets that are important to rural livelihoods is conveniently managed in terms of 'capitals', many of which already have an established place in the literature. Social capital completes this symmetry and provides a useful balance with human capital: 'where human capital resides in individuals, social capital resides in relationships' (Woolcock, 2002, p21). The various aspects of social organization subsumed under social capital are of unchallenged importance for understanding rural development; but whether they are most efficiently addressed under a single rubric is a matter of debate. There is considerable concern about the utility and purpose of the concept of social capital. Harriss and de Renzio (1997) question its logic and Fine (2001) sees the growing acceptance of social capital as evidence of the dominance of economists in debates about social development. However, our interest here is not to contribute to theoretical discussions, but rather to explore the practical consequences of various aspects of social organization in the development of agricultural technology. A brief examination of the major components of social capital will be followed by some examples from the relevant literature on agricultural development.

Social capital has come to include an exceptionally wide range of meanings. The first use of the term in contemporary literature was by Coleman (1988). It is significant that his interest involved the role of obligations, expectations, information channels and social norms in conditioning family and community support for students to continue their education, or, as the title of his article suggests, how social capital creates human capital. Woolcock and Narayan (2000) managed to unearth a very early use of the term, also related to education; in 1916 a school administrator in West Virginia wrote about social capital and its relation to community support for school performance. The term received one of its biggest boosts from Putnam's (1993) study of Italian political systems, which found that areas with higher social capital (measured by density of associations and groups) had more effective political institutions and higher levels of economic development. Social capital is described in that study as 'features of social organization, such as trust, norms and networks, that can improve the efficiency of society by facilitating coordinated actions' (Putnam, 1993, p167).

Many researchers who utilize the concept believe that social capital is as much about norms and trust as it is about concrete social organization. Grooteart and van Bastelaer (2002) distinguish between 'structural' and 'cognitive' aspects of social capital, and Krishna (2002) constructs a measure of social capital in Indian villages based both on respondents' actual group membership and on their perceptions of reciprocity and trust. There are some valuable examinations of the concept of trust (e.g. Gambetta, 1988); but it remains a difficult concept to operationalize. For the purposes of this study, we shall focus on the structural side of social capital rather than social norms. Beliefs and norms are, of course, relevant to an understanding of LEIT; but they will be discussed outside the context of social capital.

Several other dilemmas confront the definition of social capital, and these tend to echo concerns and debates from earlier literature. One question is whether social capital should emphasize formal or informal organizations. Coleman's work focused on informal networks of neighbourhood and kinship, while Putnam examined the contributions of formal clubs and associations. Formal group formation has become a common strategy for many development activities, including LEIT. Ostrom et al (1993, p220) suggest that 'grassroots local projects become part of the social capital of the future'. There is the hope (or the conceit) of those involved in development projects based on group formation that virtually any attempt at supporting organizations makes a contribution to present or future social capital. This has some relation to

Hirschmann's observation that many successes in grassroots development draw on networks created in earlier activities, manifesting a 'conservation and mutation of social energy' (cited in Bebbington and Perreault, 1999, p413). A positive interpretation would see the gradual development of experience and commitment in rural development, while a less charitable view could link such networks to the never-ending progression of development projects.

Despite the emphasis on formal groups, informal social relations are equally important for understanding information flow. The importance of informal networks was recognized long before the 'discovery' of social capital. Mitchell (1973) reviewed literature on networks from the 1950s and 1960s, illustrating their role in providing information, facilitating specific transactions and establishing norms. Although formal groups capture most attention in contemporary studies of social capital (partly because they are easier to observe and measure), informal networks certainly play a key role in the development and transmission of innovations.

A second question relates to whether social capital that provides links between groups (formal or informal) is more important than that which is exclusive to a particular group. Putnam (2000) calls these 'bridging' and 'bonding' social capital, respectively. There is recognition that while strong but exclusive groups may signal the existence of social capital, they may serve to channel or preserve privilege and perhaps foster 'antisocial capital'. The question of who is included, and who is excluded, is relevant to a consideration of group formation and rural development. Thus, bridging social capital has been attracting increased attention. On more practical grounds, there is evidence that people often gain important advantages from what Granovetter (1972) calls 'weak ties' – social links that do not exhibit the intimacy, intensity or reciprocity characteristic of interactions between members of a formal group. These instances of bridging are particularly relevant to our interests because Granovetter points out that innovations are often spread through such weak links. Rogers resurrects the earlier sociological concept of 'heterophily' (the dissimilarity between two individuals who communicate) and finds that 'heterophilous network links often connect two cliques, thus spanning two sets of socially dissimilar individuals in a system' (1995, p287). He goes on to give examples of how such links foster the diffusion of innovations.

A further, related issue is whether the concept of social capital, despite its interest in bridging, is adequate to address the importance of political relations. There is relatively little attention in the social capital literature to the development of links between local groups and those who hold power and resources. This is explored in a critique of the World Bank's (and others') endorsement of the concept of social capital by Harriss (2001). He is concerned that Putnam's relationship between social capital and political performance in Italy may provide false expectations for development activities conceived as building social capital. The argument is concerned that local associations and NGOs 'are not necessarily democratically representative organizations, nor democratically accountable, and might be attractive because they appear to offer the possibility of a kind of democracy, through "popular participation", but without the inconveniences of contestational politics' (Harriss, 2001, p9).

Political action does not go completely unnoticed in the literature on social capital. For instance, Bebbington (1999, p2032) discusses 'investments in social capital' for migrant labourers, including 'strengthening organizations that demand work safety for casual labour'. Brown and Ashman (1996, p1471) define social capital as comprising two aspects: local grassroots organizations and links 'that span differences in sector and power'. It is thus possible to include political action under the umbrella of social capital; but does this stretch the term too far to provide practical contributions?

In a study of the development performance of villages in Rajasthan, Krishna (2002) found an interaction between social capital (measured through an index that includes membership in informal networks and attitudes of cooperation and trust) and effective local leadership that linked communities with outside agencies. The emerging class of political agents described in the study is able to link villagers to state institutions and government programmes by facilitating access to information. In this case, social capital is most useful if it can take advantage of effective political brokerage. 'It is the interaction of these two variables – social capital and agency capacity – that is crucial for distinguishing between high-participating and low-participating villages' (Krishna, 2002, p169). These conclusions illustrate the limitations of seeing social capital as something that can be created for development purposes independent of the wider political context.

The nature and boundaries of social capital are thus subjects of considerable debate. A concept with such a broad range of definitions opens the door to a wide variety of interpretations. If social capital is 'the norms and networks that enable people to act collectively' (Woolcock and Narayan, 2000, p226), the term is exposed to charges of circularity. The confusion is evident in the fact that some authors use indices of social capital that include both cognitive and structural measures (i.e. assessing beliefs and membership), while others concentrate only on normative measures (such as trust) or on organizational attributes. A further issue of contention is whether it is best to examine social capital at an individual or only at a community level (or both). As Bebbington (1999, p2036) points out, social capital has become a term 'whose indicators are largely surrogate and indirect'. Thus, caution and as much precision as possible are required when examining applications of the concept of social capital to agricultural development.

Social capital and agricultural change

These problems of definition have not deterred people from making rather ambitious claims about the functions of social capital in agricultural development. One of the most noteworthy examples is a World Bank study of Tanzania that constructs an index of social capital (based on the number and characteristics of groups to which villagers belong) and finds it related to higher average household expenditure (Narayan and Pritchett, 1997). More relevant to our concerns, the study also finds that higher village social capital is related to an increased use of fertilizers, agricultural chemicals, purchased seed and agricultural credit. The study claims that the causality runs from social capital to expenditure (and not the other way) – that is, 'being part of a group makes a farmer wealthier', not 'wealthy farmers are more likely to join groups'. This claim is based on some heroic econometrics, using trust as an instrumental variable in the analysis. The assumption is that trust in others, measured by a series of questions, cannot be related to economic status, but is related to social capital. Despite the sophistication of the study, there is nothing that gives any hint regarding what aspects of social capital might be important for improving the World Bank's lending in areas such as agricultural extension. Indeed, the paper concludes that 'this research has not empirically identified any policy levers available to expand social capital or estimated the costs of creating social capital' (Narayan and Pritchett, 1997, p36). Nevertheless, the study encouraged the World Bank to pursue similar work. Elsewhere, a study in Burkina Faso includes both individual and village-level indices that are based on measures of trust, group membership and organizational type (Grootaert et al, 2002). The same positive relationships with welfare found in the Tanzania study, and the same lack of practical conclusions, are evident.

Despite significant problems with the concept of social capital, many of the aspects of social organization that it attempts to cover are undeniably important to any consideration of technical change in agriculture. A common sense approach to the components of social capital related to the uptake of agricultural innovation might well begin with elements such as the networks that promote a flow of information; the types of groups or associations that are instrumental for technology change; and the relative importance of linkages that stretch beyond individual farm communities.

If low-input technology is, indeed, information-intensive, it is important to ask where farmers might be expected to get information and to what extent local networks might facilitate the flow of indigenous knowledge, scientific principles or attitudes regarding resource conservation. In the earlier section on 'Information', some of the adoption literature that examines the importance of neighbouring farmers as sources of information was discussed. Foster and Rosenzweig's (1995, p1205) study of the adoption of modern varieties in India finds evidence of learning from others: 'farmers with experienced neighbours are significantly more profitable than those with inexperienced neighbours'. There is some evidence from other studies of the importance of kinship and other social networks for the transmission of information (and seed) of crop varieties (Tripp, 2001).

Various factors may affect the flow of information among farmers. Sumberg and Okali (1997) speak of the 'transparency of the environment' that allows (or impedes) farmers seeing and visiting each other's fields; for instance, information may travel more quickly through open irrigation tracts than between isolated forest clearings. They also found differences in farmers' willingness to share the results of their own experiments. Interest in other people's fields may be subject to social restrictions; in some cultures, attention to a neighbour's field may be taken as an indication of malign intent or witchcraft. In a study of Ghanaian pineapple growers, Conley and Udry (2001) asked farmers not only about their own practices and outputs, but also about those of other farmers. They found, perhaps not surprisingly, that farmers receive information from relatively few of their counterparts. When Sumberg and Okali (1997) asked farmers in several locations in Africa to identify local innovators, they found few farmers willing or able to point to others who were known for experimenting; and in most cases those who were named were not outstanding innovators.

Although information flow among farmers within a community may be less than perfect, there is a good argument to be made for the potential importance of expanding informal contacts among farmers across villages and even across regions and countries. This helps to explain the growing popularity of group visits organized by development projects to see innovations or exchange views. Expanding farmers' range of contacts is a way of diffusing information and ideas and giving farmers confidence by showing them what their counterparts in other areas have been able to do.

A subject that has received more attention than the role of informal networks is the degree to which organized groups serve to further agricultural development. Interest in this subject predates the emergence of the social capital concept. Anthropologists have described how kinship organizations (Netting, 1968) and other organizations such as church groups (Long, 1968) facilitate access to labour and resources for agricultural activities. Wade (1988) provides a detailed account of the conditions and incentives that encourage local village organizations to manage and regulate access to irrigation in India. The basis for group formation may vary over time; in their analysis of soil fertility management in southern Ethiopia, Konde et al (2001) describe how traditional work parties have been replaced by church groups as sources of labour and support.

Social capital is sometimes seen as a manifestation of the totality of associations and networks. This involves the conviction that social capital is something greater than the sum of group memberships. This conception makes social capital a quality of communities or regions, rather than a resource of individuals, as Putnam's (1993) study emphasized. Communities can certainly be differentiated with respect to their organization and past experience. Boyd and Slaymaker (2000) discuss the importance of 'social cohesion' in determining capacities to take on soil and water conservation activities. Humphries et al (1999) found that differences in community experience in Honduras made a big difference to their capacity to organize agricultural experimentation groups. Whether such differences are best described in terms of social capital or more simply in terms of past history of interaction with external agencies is open to question.

But even if this view of social capital as a community resource has validity, individuals certainly differ in their ability to take advantage of it. Grootaert et al (2002) point out that poorer households in Burkina Faso belong to fewer associations than the rich, but hold that this limited social capital is of relatively greater importance to marginal households. However, it should not be necessary to warn against the myth of community homogeneity or to point out that group activities do not necessarily guarantee equitable access to assets such as social capital or to new technology.

Group formation

The recent interest in social capital in the agricultural development literature follows a wider trend of ignoring indigenous organizations and networks and concentrating on the formation of new groups. This emphasis on group formation signifies an evolution of the notion of social capital in the wider literature. The original concept focused on an individual's spontaneous utilization of previously established social networks for other purposes (for example, Coleman and education). This was supplanted by interest in the more general impact of an environment of social networks on development (for instance, Putnam and political advance), while current interest is shifting to outside interventions that create groups to address a deficiency in local social capital. This more purposive approach is understandable from a development perspective but is not always consonant with an interest in indigenous, grassroots initiatives.

One of the most thorough expressions of this faith in group formation is found in a recent article by Pretty and Ward (2001). The article includes an inventory of over 400,000 group activities related to environmental betterment, representing 'advances in social capital creation' (Pretty and Ward, 2001, p214). Almost all of them are the products of development projects. The groups considered exhibit quite different characteristics. At one extreme are examples such as irrigation users' groups, which may be expected to achieve a degree of permanence; at the other extreme are microfinance groups (accounting for two-thirds of the total inventory), which are the generally shorter-lived products of development projects. Indeed, the study is more useful as a summary of current development agency fashion than as an assessment of grassroots activity. The article gives the impression that externally created groups are needed to replace the local institutions that previous colonial administrations and top-down development strategies destroyed. The authors propose an evolutionary scheme for the development and sustainability of such groups.

Farmer groups, of course, have a long history of contributions to agricultural advance. But faith in the ability of development projects to create viable local organizations may be tempered by examination of the conditions and rationale for independent farmer group success. For example, the number of farmer groups in England expanded greatly, beginning in the 18th and continuing into the 19th century. The most rapid expansion corresponded to a time of unprecedented prosperity for grain and livestock farmers, who were eager to acquire information and access to the latest innovations (Fox, 1979). The roles of economic incentives and local initiative were paramount. 'Those that tended to succeed were established by groups of local farmers themselves, rather than imposed by outsiders, or had access to outside financial or patronage support' (Pretty, 1991, p137).

This rush towards addressing agricultural development challenges by the formation of farmer groups certainly has some practical limits. For instance, the majority of efforts at developing village seed producer groups and converting them into commercial enterprises have been failures (Tripp, 2001). Böhringer et al (2003) examine tree nurseries established by groups and individuals, usually in the context of project support, in several countries in Africa.

They show that individual nurseries are usually much more productive, and that group nurseries often suffer from disputes over responsibilities and management. However, many of the successful individuals were 'graduates' of nursery groups, where they had learned their skills.

In addition, group development activities may, at times, be much less equitable than is apparent. For instance, there is an emerging literature on microfinance that questions the equitability of some of these schemes and the groups that they engender (see, for example, Mayoux, 2001). Gugerty and Kremer (2002) go so far as to argue that some bottom-up rural development activities organized through groups may offer more opportunities for resource capture by an elite than do traditional top-down activities. A study in Kenya examining the impact of projects promoting agroforestry innovations found that many villagers complained that group formation often resulted in domination by better-off farmers (Place et al, 2003).

There are many good reasons for promoting group activity in rural development, but some objectivity is required. A visit to any village that receives attention from development projects or NGOs is likely to uncover any number of groups, committees and clubs. A passage from a study in Zambia may be a useful counterbalance:

> Many villagers understand that outsiders have control over goods and services, otherwise inaccessible. Some become adept at manipulating those outsiders, especially through the language of 'participation'... A manifestation of the 'development effect' is the formation of clubs. Of course, diverse groupings are made for a wide range of productive and distributive purposes. They may be, for the farmers concerned, the most effective means of using restricted resources. However, where the objective or rationale for group formation is at least in part influenced by outsiders, outcomes may be less positive (Harrison, 1996, p276).

Summary

This chapter has reviewed literature related to the roles of labour and information in shaping farmers' choices of new technology. These two factors are often cited as being particularly important for LEIT, which may require the reorganization of labour and the use of new information to help substitute the management of local resources for dependence upon external inputs. The literature is drawn from a range of sources, but includes formal adoption studies and recent literature related to the sustainable livelihood approach and, in particular, social capital.

The labour component of any agricultural technology is a key determinant of a farmer's decision to consider a particular innovation. This is not to say that farmers necessarily reject technologies that require more labour; but they need to be convinced of the benefits and to be able to gain access to that labour, taking account of seasonal demands, the social organization of village labour and the local labour market. Increased farm labour demands are often met by hiring labour. The availability of household or hired labour depends to a considerable extent upon non-farm labour opportunities and decisions regarding household livelihood strategies. These decisions, in turn, are conditioned by information about the performance and requirements of new technology and perceptions of the long-term viability of the farm.

Information is important for all types of farming, and it may not be accurate to characterize LEIT as being particularly information-intensive. Information can be obtained from external sources, such as the extension service; but the emphasis in the promotion of LEIT is often on local flows of information. Farmers may make judgements about new technology based on their own experience or in reference to the experience of other farmers. These two information sources correspond roughly to the difference between human and social capital.

Human capital has a fairly long history in the study of agricultural technology adoption, although most of the formal studies have concentrated on formal education as a determinant of adoption. Other aspects of human capital, such as the capacity to innovate or access to ITK, are of equal interest for technology adoption but have received less formal attention.

The subject of social capital occupied some space in this review because of the potential importance of social organization for the diffusion of LEIT, and because the concept of social capital has become such a prominent topic in discussions of rural development. The theme has featured in a number of recent studies, although there is, as yet, little research that formally addresses the relationship between social capital and the diffusion of agricultural innovations. Nevertheless, there is some literature indicating the importance (and the limitations) of local networks and formal organizations for promoting the utilization of agricultural technology. Much recent development effort in promoting appropriate agricultural practices has been linked to social capital through an emphasis on group formation. Less emphasis to date has been given to the development of 'bridging social capital' that would link such groups with outside agencies.

The observations in this chapter have been motivated by an interest in two factors (labour and information) that are particularly important for LEIT, and the assets (human and social capital) used to manage those factors. The literature reviewed made little reference to particular agricultural technologies. The next chapter examines specific experiences with the utilization of LEIT.

4

The Impact of LEIT: Evidence from the Literature

Robert Tripp

Introduction

Before examining the evidence from the three case studies, it will be useful to review what the existing literature tells us about the diffusion and utilization of low external-input technology (LEIT). As discussed in Chapter 1, there are relatively few studies that analyse the adoption of this type of technology. This is partly because most efforts at introducing and promoting LEIT are comparatively recent, and experience regarding the adoption of the technology is only beginning to emerge. Another factor possibly explaining the paucity of analysis on the uptake of LEIT is the assumption that this type of technology is naturally appropriate for lower-resource farming households and, hence, does not require the same scrutiny regarding the distribution of benefits that conventional agricultural technologies often attract. This chapter attempts to summarize the available literature on the utilization of LEIT and, in particular, to examine what is known about differences in experience within and between farming communities.

Although much of LEIT is based on traditional crop management practices, this review is interested primarily in the experience of developing low-input strategies that are new to particular farming systems and encouraging farmers to make better use of such technology. Thus, we shall examine the fate of technologies that, while largely based on locally available resources, require farmers to learn new skills (or refresh old ones) and make changes in their normal farming practices. While not attempting an exhaustive assessment of the literature, the chapter examines as wide a range of experience as possible.

There are a number of factors that may explain the patterns of utilization and diffusion of any technology. One of the primary issues is technical feasibility. Farm management innovations that look promising 'on the drawing board' (in an experimental station or a demonstration) often fail when introduced to farmers' fields because of physical or biological interactions that were not foreseen. LEIT is not immune to this type of failure, and there are a number of instances where promising LEIT has simply not performed to expectations in the agronomic environment in which farmers operate. Any innovation should offer clear technical advantages for farmers, and this has not always been the case with LEIT. Reviews of alley cropping (Carter, 1995; Dvorák, 1996) point to a number of problems, including lower than expected biomass production and poor performance of some species. Cramb et al. (2000) describe difficulties with hedgerows in the Philippines, including maintenance problems and the adverse effects of waterlogging in heavy soils. Stocking (1996) provides cautionary examples regarding the unintended effects of apparently logical soil conservation remedies. Integrated pest management (IPM) faces many technical challenges that render a number of current techniques unreliable (Jeger, 2000; Way and van Emden, 2000).

However, our interest in this study is concentrated on the human and social capital correlates of technology utilization rather than on technical performance. We shall therefore concentrate on examples of LEIT that exhibit an acceptable degree of technical feasibility, and will not review studies that examine the extent to which variability in technical performance affects farmers' experience with an innovation, although this is an important theme in the adoption literature for LEIT.

Thus, this chapter concentrates on cases that allow some understanding of who is able to take advantage of (technically appropriate) LEIT and the impacts and implications of resulting utilization patterns. The discussion begins by briefly reviewing a number of case studies that give an indication of the equity implications of LEIT. The review is particularly interested in asking if this type of technology has been effective in reaching the poorer households in farming communities and providing options for more marginal farming areas. Unfortunately, the literature does not provide anything close to a uniformly positive response.

In order to understand this uneven experience, the rest of the chapter examines several factors that may help to explain the variable uptake of LEIT. General economic incentives for farming and investment in technology play a role. Particularly important is the constellation of factors often categorized as 'labour' that involve complex decisions about how members of farming households spend their time. Human capital also plays an important role, and the review examines how farmers' education, skills and attitudes influence (and are influenced by) the progress of LEIT. Attention is also given to the occasionally important relationship with larger political movements that become associated with resource-conserving technology. The final sections of the chapter examine the interplay of social organization and technology utilization and its relevance to the flow of information and support for the utilization of LEIT.

The poverty focus of low external-input technology (LEIT)

A major interest of this study is to understand the equity implications of policies supporting LEIT. There is the widespread hope that such technology is capable of ameliorating some of the negative distributional effects of conventional agricultural development, and that it offers a significant opportunity for delivering benefits to relatively disadvantaged farm households. Although global conclusions regarding the impact of such a broad class of technologies on the extent and distribution of farm-level poverty are not feasible, this section reviews some evidence on the patterns of adopting LEIT. Much of the rest of the chapter then reviews the literature on the relationship between specific aspects of farmer resources and the uptake of these technologies.

Marginal environments

An examination of the equity implications of any agricultural technology can address differences within communities, as well as differences between communities or regions. For instance, the considerable literature concerned with the outcomes of the Green Revolution encompasses both of these aspects (e.g. Lipton with Longhurst, 1989; Freebairn, 1995). One of the principal concerns about the Green Revolution is that it was usually implemented in areas with above-average resources, such as irrigation and good access to markets. The concentration of effort in these favoured environments raises questions regarding the equity impact of technology development. Successful technology adoption by certain communities or regions (involving some or all of the farmers) can come at the expense of attention to farmers in more marginal environments who are less able to take advantage of the new technology, thereby widening inter-regional differences. A review of the early literature on the Green Revolution showed that in many cases the adoption of modern crop varieties and associated technology in better-endowed regions widened the gap between those areas and less well-endowed ones (Lipton with Longhurst, 1989). An examination of more recent experience with seed-fertilizer technology in south India showed that, although in irrigated areas small farmers achieved higher proportional increases in household income than large farmers over a ten-year period, farmers in areas without irrigation made only small proportional income gains (Hazell and Ramasamy, 1991).

This regional inequality in relative gains from technology is the basis for a concern about the investment of resources in agricultural research that asks whether marginal production environments deserve more attention (Renkow, 2000; Ruben and Pender, 2004). During recent years, this inequality has also become an important element in discussions about the relative importance of agriculture and agricultural technology as potential contributors to sustainable rural livelihoods in marginal areas, and has served to focus attention on alternative, non-agricultural income sources (Berdegué and Escobar, 2002). The interaction between these non-farm opportunities and the incentives for the utilization of new agricultural technology is an important factor in understanding the career of LEIT.

The call for more investment in agricultural technology for marginal areas is often made by supporters of LEIT, who argue that such technology is particularly appropriate for environments neglected by previous efforts at technology generation and that it can bring real gains in productivity and welfare to these areas. This chapter is not the place to attempt an assessment of the geographical targeting of LEIT efforts; but clearly many of the most important LEIT programmes are, indeed, aimed at more marginal farming environments. This is partly because of the nature of some LEIT (that seeks to ameliorate the effects of poor soils, inadequate rainfall and other environmental challenges) and partly because of the commitment to poverty reduction of many of those who work on LEIT.

Of course, the supporters of LEIT do not argue that these technologies are only suitable for marginal environments. Indeed, one of the most important arguments is the ability of LEIT to substitute for high dependence upon external inputs, a chararacteristic of more favoured environments. For instance, much of the work on IPM takes place in irrigated rice systems with farmers who are well connected to input and output markets. A recent analysis of 208 projects that promoted 'agricultural sustainability' (focusing on, but not limited to, LEIT) involved nearly 9 million farmers and 29 million hectares. Although only 4 per cent of these were classified as large farmers, they accounted for more than 70 per cent of the total area of adoption, mostly large commercial farmers in Latin America who were practising conservation tillage (Pretty et al, 2003).

Intra-community differences

Even when efforts are targeted at marginal areas, questions may be asked about the intra-community distribution of any benefits. Many studies on the impact of the Green Revolution examine the degree to which larger or wealthier farmers within a village were able to take better advantage of the seed-fertilizer technology, and, if they did, whether their less well-resourced neighbours were permanently excluded or were eventually able to catch up. The same types of questions can be asked about the experience with LEIT, although the literature available is in no sense equivalent to that for the adoption of external inputs. There is hope that LEIT is more scale neutral (or even poor friendly) than many external-input technologies; but that assertion must be demonstrated in particular cases.

A preliminary examination of the somewhat limited literature available provides more examples in which LEIT adoption favours relatively wealthier farmers than those in which poorer households have at least equivalent access to the innovations. The use of a cover crop in northern Honduras (spontaneously adopted, with no external promotion) tends to be associated with farmers who grow larger amounts of maize and are more commercially oriented (Neill and Lee, 2001). A study of farming practices in an area of western Kenya showed that wealthier households are more likely to use both external inputs (hybrid maize, inorganic fertilizer) and low-input soil management techniques (fallows, compost, terraces) (Crowley and Carter, 2000).

In Niger, the introduction of planting pits has helped to rehabilitate degraded land and has contributed to an emerging land market; but the purchasers are concentrated among a rural elite (Hassane et al, 2000). In Burkina Faso, the poor are less likely to take advantage of this technology because of its labour requirements (Kaboré and Reij, 2004). The practice of mulching (cutting grass to carry and spread on fields) in a village in Burkina Faso is more common among larger farms, which are more likely to have donkey carts to transport the grass, and among those farms with a higher available work force (Slingerland and Masdewel, 1996). The initial experience with the system of rice intensification (SRI) in Madagascar indicates that adopters are more likely to be surplus rather than deficit rice producers, with more land and often with better off-farm sources of income (Moser and Barrett, 2003). A project that successfully introduced contour hedgerow intercropping in an area of eastern Indonesia drew a preponderance of its participants from wealthier households (Wiersum, 1994). These were households that had sufficient land for subsistence and could afford any temporary decrease in production while the agroforestry intercrop was being established. Kelly et al (2002) discuss the challenges for a regional development programme in Niger to extend its natural resource management programme (featuring erosion control and organic soil amendments) beyond the better-off farmers in higher-rainfall villages.

Similarly, Sanders (2000) found that it is wealthier villages in China that are more likely to adopt 'ecological agriculture' (building on traditional Chinese organic production techniques). A movement towards 'restorative agriculture' in the US during the early 19th century (featuring a shift from extensive cultivation towards more careful soil management and the use of manure) was distinguished by the fact that 'the majority of improving farmers held a fortune somewhere above middling, including merchant squires of great wealth... Those who incorporated restorative methods almost always lived close enough to market towns to turn surplus into cash' (Stoll, 2002, p28).

These few pieces of evidence of unequal access to the benefits of LEIT are far from conclusive; but they certainly indicate that LEIT is not immune from the distributional biases that often characterize other types of agricultural technology. The evidence is not universally negative, and there are studies that show more complex relationships between wealth and the uptake of the technology. For instance, a study in the Philippines shows contrasting patterns with respect to the adoption of soil conservation measures; in some villages it is the poorest households which are most likely to adopt, while in others it is the wealthiest (Lumley, 1997). A study of another soil conservation programme in the Philippines showed that it tended to benefit a 'clique of "yeoman farmers"' (Brown and Korte, 1997, p14), but not the village elite.

Part of the problem may simply be the fact that wealthier households are likely to be early adopters of any technology. There is much literature from the Green Revolution showing how smaller farmers soon caught up with their larger neighbours (e.g. Lipton with Longhurst, 1989), and a similar pattern may be relevant for LEIT. Franzel (1999) discusses the fact that while wealthier households in Kenya and Zambia tend to be the first to take advantage of improved fallows, there is no obvious barrier to their use by poorer households. In Zambia, although wealthier farmers adopted the improved fallows to a larger extent, there is no difference between wealth groups in the propensity to continue with the technology once initiated (Franzel et al, 2002).

Another part of the explanation for the less than equitable experience of many LEIT efforts may be the nature of project management. Projects often begin in areas, and with farmers, that are more accessible. For instance, the early experience of farmer field schools (FFS) in introducing IPM in Indonesia concentrated on 'the better informed and more affluent farmers' (Roling and van de Fliert, 1994, p103). Fakih et al (2003) discuss the problems of overcoming the biases in the way that extension agents select participants for IPM programmes. A successful project that introduced soil conservation techniques to highland Ecuadorian farmers was careful to select communities that had good internal organization and low levels of migration (Winters et al, 2004). Such bias can presumably be overcome as projects gain experience with the techniques that they are promoting and begin to focus on a broader range of clientele; but the dangers are considerable. A study of the impact of agroforestry innovations in Kenya found no relationship between wealth and adoption in the pilot villages that received extra attention for reaching disadvantaged groups, but did find a bias towards better-off farmers in villages included in a subsequent round of less intensive promotion (Place et al, 2003).

Other project management biases may be more difficult to overcome. LEIT activities may be susceptible to capture by local elites. Projects promoting the construction of stone bunds in Burkina Faso relied upon the formation of village groups:

> Influential village members, such as the better-off, local chiefs and the so-called enlightened (those allegedly well versed in modern ways, often returned migrants), tend to dominate these groups. While there is usually an atmosphere of free discussion, it is true that the dominant ones are able to rely on the group to help them develop their own farmland. The less well-off do participate in the activities of the group, although they rarely benefit directly. Quite often, their own farms are neglected (Atampugre, 1993, p106).

Kerr (2002) reviews the results of India's huge investment in watershed development projects that attempt to increase rain-fed agricultural production and conserve natural resources through various improvements in land management. His conclusion is that:

> ... the projects most successful in achieving conservation and productivity benefits also had the strongest evidence of skewed distribution of benefits toward larger landholders ... watershed development often asks the poorest, most vulnerable people to provide a valuable environmental service to the wealthiest landowners (Kerr, 2002, p1398).

He adds that more research is needed to see if indirect benefits, such as increased agricultural employment, may lessen the gap represented in these findings.

Thus, even when projects specifically target LEIT at marginal production environments, there is still the likelihood that the more affluent farmers will be better able to take advantage of the innovations, at least initially and, perhaps, permanently. In this respect, there may be less difference between LEIT and other technology than is generally recognized. Place et al (2002b, p281) conclude that natural resource management technologies, 'like agricultural technologies more generally, fail to be adopted by women farmers and poor farmers at the same rate as male farmers who enjoy greater wealth, education and socio-economic power'. When these sorts of biases are identified, they do not imply that the particular technology is inappropriate or should be abandoned. But the results argue for more effort to understand patterns of uptake and distribution and to work towards improving targeting, project management, policies or the technologies themselves so that a wider range of farmers can take advantage of the innovations. The uneven spread of LEIT may also point to limitations in an agricultural technology-led strategy for poverty alleviation in marginal areas.

The rest of this chapter examines some of the factors that explain these observed patterns of uptake for LEIT and helps us to understand why some experiences with LEIT have been less poor-friendly than expected.

Economic incentives

A farmer will not invest in a new technology unless it offers a reasonable economic return. Calculating those returns, especially when the investment implies a redeployment of household labour or the acquisition of new skills, or when the results may only become obvious after several seasons, can be a tricky business. But before examining the ways in which the management of human and social capital influences the acceptability of LEIT, a few words should be said about the general economic environment into which LEIT might be introduced. In most cases, agricultural technology finds a better reception where farmers have secure access to their land, where farming is remunerative and where agricultural markets perform adequately. There is no evidence in the literature to indicate that LEIT is an exception to these patterns.

The economic environment

Much has been written about the importance of land tenure and security for a commitment to conservation strategies, especially when it involves a long-term investment or pay-offs. Although the basic logic is unassailable, it is important to note the variable role of land tenure in providing incentives for LEIT. For some technologies, such as IPM, the farmer can achieve benefits in a single season, and land tenure may be relatively unimportant. On the other hand, for making investments in long-term farm improvement, such as soil management, secure tenure is an obvious advantage. Wiersum (1994) discusses how some households in Indonesia participated in a soil conservation project specifically as a way of gaining secure tenure. But the situation is not necessarily straightforward, as several studies have shown. In one area of the Philippines, tenant farmers invested in soil conservation because they had fairly secure and long-standing access arrangements, while in another area less secure tenancy arrangements were a disincentive to adopting these technologies (Cramb et al, 2000). Lutz et al (1994) found, in a review of Latin American soil conservation projects, that insecurity of tenure was not necessarily a disincentive for investment, as long as the conservation measures had short payback periods. A study in Haiti revealed that the complex and fluid local tenure system did not strictly predetermine farmers' investments in various soil conservation technologies (Smucker et al, 2002). Francis (1987) discusses how differences in tenure and customary land access affect farmers' incentives for the utilization of alley-cropping techniques among different ethnic groups in Nigeria.

The profitability of farming is an important incentive for adopting any agricultural technology. Cramb et al (2000) compare the experiences of several soil conservation efforts in the Philippines and find that adoption is highest in an area where proximity to large urban markets gives farmers an incentive to conserve their soil. A project that worked with farmers in Kenya and Uganda to understand soil fertility problems and to test and adapt various low-input interventions concluded that the current socio-economic environment was not conducive to investments in any type of soil nutrient management (de Jager et al, 2004). Similarly, Orr (2003) argues that one of the reasons for the lack of success of most IPM efforts in sub-Saharan Africa is the absence of a dynamic commercial agriculture that would stimulate investment in farming and elicit demand for this type of technology.

Cash crops often appear as an important factor in the decision to adopt LEIT. A series of African case studies found that most adoption of soil and water conservation technologies was related to the cultivation of high-value crops (Boyd and Slaymaker, 2000). Farmers in remote areas of northern Morocco have found that the high profits from (illegal) marijuana cultivation provide sufficient incentives to maintain and improve traditional soil conservation structures such as stone bunds and to invest in soil conservation on newly cleared land (Chaker et al, 1996). The spontaneous adoption of conservation farming (i.e. planting pits) in Zambia is higher among cotton farmers than among those who grow only food crops (Haggblade and Tembo, 2003). There is evidence that many instances of the adoption of soil conservation measures (particularly terracing) for food crops in parts of eastern Kenya are related to a village's proximity to markets and, in some cases, to the ability to invest windfall profits from high coffee prices in labour for terrace construction (Zaal and Oostendorp, 2002). Place et al (2002a) find that biomass transfer of leguminous shrubs in western Kenya is most often applied to small fields and particularly for higher-value vegetables, rather than for subsistence crops. In northern Honduras, the use of cover crops for maize is more common among those farmers who grow high-value crops such as citrus or coffee (Neill and Lee, 2001).

The opportunity for expanded commercial production may also be an incentive for the adoption of LEIT. Wiersum (1994) found that those Indonesian farmers adopting contour hedgerows were better able to afford the extra labour inputs and looked forward to converting the newly terraced land to cash-crop production. Farmers in the Philippines incorporated vegetative strips as a soil conservation measure in their farms, but then managed the conversion from grass strips to the establishment of contours with fruit or timber trees that were of much greater economic value (Mercado et al, 2001).

However, markets can work for or against LEIT. Sain and Barreto (1996) compare the variable experiences of several areas in El Salvador with conservation tillage for maize (using herbicide). One of the most significant factors explaining adoption in one area and rejection in a nearby one is the difference in the value of maize stover (used as mulch). Farmers in the area with lower animal ownership and the absence of a market for stover are more likely to conserve sufficient crop residues for mulch because of their low value. In areas where the stover has a higher value, it is more difficult to persuade farmers to use it for soil conservation. Similarly, a successful project that used low-input strategies to restore soil fertility in hillside Honduras was based, in part, on the ready availability of chicken manure from nearby commercial poultry operations (Bunch, 2002). Increasing demand for manure led to a rise in prices and eventually saw farmers shifting towards mineral fertilizers. Other examples of competing uses for farm by-products include the importance of manure or stover for fuel in some farming systems, limiting their use as soil fertility amendments.

Changes in the local economy may also work against certain technologies. Neill and Lee (2001) describe how changes in land markets and the expansion of the cattle industry and road infrastructure in Honduras made the use of cover crops on hillside maize less attractive as farmers found alternative uses for land and labour. Cramb et al (2000) discuss the abandonment of hedgerows in one area of the Philippines, in part due to a shift away from the crop (upland rice) that formed the basis for the practice.

External support

Some assurance of economic performance may be part of project design for the introduction of LEIT. Winters et al (1998) analyse the success of a programme run by an non-governmental organization (NGO) in Ecuador that offered farmers information and some incentives for a wide range of conservation techniques. The considerable success that was noted was, in part, due to attention to ensuring that short-term economic gains were evident. The added incentives for adoption provided by the project were not payments for conservation measures, but rather access to complementary inputs. The (limited) adoption of alley cropping in some areas of West Africa may have been related more to project provision of inputs and veterinary services than to the actual performance of the technology (Douthwaite et al, 2002). The significant increase, and subsequent decline, in the use of velvet bean (*Mucuna pruriens*) as a cover crop in Benin was related to the fact that the project initially purchased seed from early adopters to use in further promotion and then stopped its purchases when supplies were sufficient (Douthwaite et al, 2002).

The role of economic incentives for the diffusion of LEIT, and the degree to which they can be provided without external subsidies, is an important issue. McDonald and Brown (2000) suggest that soil and water conservation technologies must either provide immediate, tangible benefits to farmers or be supported through some sort of incentive system. Hellin and Schrader (2003), on the other hand, summarize studies from four sites in Central America showing that although direct incentives motivated farmers to establish soil conservation structures, once the incentives were removed further structures were not established and those remaining were poorly maintained. Barbier (1990) discusses the experience of soil conservation projects in Java. Poorer farmers growing subsistence crops may be aware of soil erosion but unable to afford control measures; even wealthier farmers growing highly erosive but profitable crops may not change their practices until the effects of poor land management become obvious.

There is the question of whether LEIT can lead the way to more productive and profitable agriculture, or whether such a transformation must await policy or other changes that provide incentives for farmers to invest in new technology. Bunch (2002) discusses the evolution of technology use in Guatemalan and Honduran communities that began with experimentation with a few basic soil conservation and regeneration innovations for subsistence crops. Success with these methods eventually led to much more active participation in commercial agriculture opportunities, such as vegetable growing (see Chapter 5). In this case, it would seem that LEIT can play an important role for broader agricultural development. Murray (1994, pp142–143), on the other hand, believes that market development is primary:

> … without some catalysing technical or economic change that dramatically alters the lethargic cost-benefit regimes of traditional agrarian systems, it is unlikely that new soil conservation will spread, either through spontaneous local development or through project-mediated promotion … innovations in soil conservation should not be expected to function as the catalysing technical innovation that catapults a dormant agrarian system into forward movement. Rather, soil conservation as a domain is most effectively introduced as an ancillary adjunct to inputs more capable of generating increments in short-term productivity.

Economic incentives: Summary

In summary, there is little indication from the studies reviewed that farmers subject most classes of LEIT to different criteria from those they use when considering other types of technology. Acceptable economic performance is a prerequisite, and such performance is more likely to be realized in an environment where policies and infrastructure support a dynamic agricultural sector. Thus, it should not be surprising that patterns of LEIT adoption are not very different from those of technologies relying upon external inputs, and that wealthier, better-connected farmers are often the first to adopt LEIT. Farmers' assessment of any type of technology is not necessarily straightforward, however. LEIT may present particular challenges, because its utilization often implies some reorganization of farm resources and farming skills. The rest of this chapter examines some of the most important factors that can influence these decisions, beginning with the issue of labour investment.

Labour

One of the most commonly cited constraints to the utilization of LEIT is labour. However, as discussed in Chapter 2, it is inaccurate to characterize LEIT as necessarily labour-intensive. Some examples of LEIT require no more labour than the farmer's present practice, and some types (such as certain variants of conservation tillage) are attractive precisely because they save labour. But it remains true that the success of LEIT is often dependent upon the efficient organization of labour supply. In some cases, the crucial factor is an initial investment in labour for the establishment of LEIT (such as a soil conservation measure); once established, the labour requirements then fall, sometimes below those of conventional management. In a number of instances, however, LEIT does imply a permanent increase in labour as management of local resources substitutes for purchased inputs. In other cases, the crucial factor is not necessarily the investment in physical labour, but rather the time for learning new skills that can be applied to the farm or for monitoring performance.

This section includes examples of these various types of time and labour investment and reviews how they interact with patterns of LEIT utilization. It concludes by examining experience regarding the nature of labour investment in LEIT, particularly the choice between household and hired labour, and the role of off-farm income.

Farm labour and LEIT

There are instances in which the initial labour investment required for LEIT is simply not available to the households that might otherwise take advantage of the technology. In Baringo District, Kenya, the first year of establishing contour ridges was estimated to require 180 hours of labour per half hectare. Adoption of the technique was low, despite a number of food-for-work programmes, because the investment had to be made at the end of the dry season, which is a period of high labour demand for agro-pastoralists (Hogg, 1988). In Burkina Faso, stone bunds contribute to soil and water conservation, promoting higher yields and, eventually, higher returns to labour. Nevertheless, 48 per cent of women involved in their construction claimed that the bunds added to their workload and only 12 per cent said the bunds lightened their work (Atampugre, 1993). In his generally positive assessment of the potential of SRI in Madagascar, Uphoff (2002c) explains that adoption is most difficult for the poorest farmers, who do not have the necessary labour at the beginning of the season.

It is not only the absolute amount of labour required, but also the timing of the labour demands that can affect the uptake of LEIT. One reason suggested to explain the fact that tied ridging has been adopted to only a very slight extent in the Sahel, despite considerable promotion, while the construction of planting pits has proved somewhat more popular is that the former technique must be carried out during land preparation, when labour demands are high, while the planting pits are usually dug during the dry season, when more labour is available (Ouedraogo and Kaboré, 1996).

In addition to those cases where LEIT requires a high initial labour investment, many studies confirm the importance of routine labour demands as a determinant of LEIT adoption. A number of observers (e.g. Morse and Buhler, 1997) have remarked on the problems that some IPM methods impose on farm labour patterns; for instance, farmers often find they do not have time to devote to scouting for pests. Carter's (1995) review of alley farming points to its high labour requirements at various times in the cropping calendar as one of the major factors explaining the technology's very limited success. Farmers in Burkina Faso have difficulty collecting and transporting organic fertilizer during the dry season when many young men are away as labour migrants (Dembélé et al, 2001). Winters et al (1998) found that the adoption of soil conservation techniques promoted in Ecuador was strongly correlated with the amount of male labour available to the household. Howard-Borjas and Jansen (2000) review the labour implications of organic agriculture in Europe and find that these techniques usually imply significantly higher labour investment, much of which is met by employing seasonal workers from other countries.

The role of labour in the uptake of LEIT is illustrated by an analysis of the use of green manures in irrigated rice production in Asia (Ali, 1999). Much of the behaviour of rice farmers with regard to green manure options can be explained by looking at the relationship between labour costs and fertilizer prices in individual countries. Green manures and fertilizer are alternative sources of nitrogen, and the cost of green manures is mostly dependent upon the labour for their management. When the ratio of labour cost to fertilizer price is above a certain level, farmers abandon the use of green manure. Even when fertilizer is priced at competitive rather than subsidized levels, the rising cost of labour indicates that in many Asian countries green manures are an economically unattractive option.

Even when labour availability *per se* is not an impediment to the uptake of LEIT, there are a number of instances where the new practices require additional skills. The skills may be required for the one-time application of a particular technique (such as the establishment of contour ridges with an A-frame); but often they are necessary each season in order to make adjustments and adaptations, and to promote further innovation. These skills may require significant time to acquire, and in this sense may be counted as 'labour' costs. Pretty sees them as part of 'transition costs' – that is, investment for transition from conventional to alternative agriculture:

> Lack of information and management skills is, therefore, a major barrier to the adoption of sustainable agriculture. During the transition period, farmers must experiment more and so incur the costs of making mistakes, as well as acquiring new knowledge and information (Pretty, 1995, p96).

The importance of gaining experience and familiarity with a technology is crucial in shaping perceptions of labour demands. Farmers in southern Mexico who had recently been introduced to the use of velvet beans as a cover crop for maize cultivation were asked to respond to a set of drawings depicting possible advantages and disadvantages of the technology. More than a quarter of the sample said that increased labour was one of the technology's disadvantages (Soule, 1997). In contrast, when farmers in northern Honduras who had long experience with the technology responded to the same questions, they did not mention labour when discussing the advantages (such as the contribution to soil fertility) and disadvantages (such as the cover crop's tendency to attract rats and snakes) (Buckles et al, 1998).

Another example of the importance of experience for shaping perceptions is provided by an IPM programme among large-scale tomato growers in California. Although many of the non-adopters were uncertain about the performance and effects of IPM techniques, those who offered opinions tended to feel that the techniques were more difficult, time-consuming and expensive than did those farmers who actually had experience with the methods (Grieshop et al, 1988). There was also a strong tendency for farmers who used IPM for other crops to adopt the IPM programme for tomatoes. Even in this quite sophisticated farming environment, the opportunity for gaining experience with new techniques is still important for being able to assess their labour implications accurately.

The time to engage in learning and experimentation is less likely to be available to poorer members of the community. In establishing farmer field schools for IPM in Zanzibar:

> ... [it was] necessary to work with groups of farmers who were willing to learn, able to experiment, had enough flexibility to change and were prepared to commit themselves for one or more seasons. This automatically excluded the poorest farmers who had little physical and financial buffer for experimentation (Bruin and Meerman, 2001, p67).

Some IPM programmes find that women (who are not the principal decision-makers for crop management) represent the household in farmer field schools when men are engaged in other types of work (e.g. Rola et al, 2002).

The techniques of SRI not only require more labour, but also new skills. Moser and Barrett (2003) point out that the labour requirements of SRI cannot easily be fulfilled with hired labour because the unfamiliar tasks associated with the system need close supervision. One observer suggests that realizing the potential of SRI will depend upon 'agricultural professionalism' in mastering the various skills required (Stoop, 2003). Similarly, the adoption of conservation farming (CF) in Zambia requires advance planning and careful execution of tasks:

> It requires a change of thinking about farm management under which the dry season becomes no longer a time primarily reserved

for beer parties and socializing, but rather an opportunity for serious land preparation work. Anecdotal evidence from our field interviews suggests that retired school teachers, draftsmen and accountants make good CF farmers (Haggblade and Tembo, 2003, p39).

The time and skill requirements of many LEIT techniques help to explain some of the patterns of utilization observed in the literature and the degree to which the technology is available to poorer households. Once again, we need to be reminded that 'labour' is not a single, undifferentiated factor. We need to consider the timing of the labour investment, its quality and whether labour costs can be expected to decrease with time (because of learning or because, once the technology is established, it requires less labour). But it should be obvious that a technology that requires 'only household labour' is not necessarily adapted to the needs of the poor.

Off-farm labour and LEIT

It is not only seasonal labour bottlenecks that may make LEIT difficult for poorer households. The opportunity costs of household labour are also important. Pender and Kerr (1998) compare the experience of indigenous soil conservation techniques for lower-caste residents of two villages in India. In one village, most of these people are involved in migratory wage labour during the dry season, when conservation measures such as stone drains and grass strips are established; as a result, their utilization of such techniques is quite low. In the other village, the lower-caste members are less likely to migrate (and less likely to be able to take advantage of local government employment schemes). With lower opportunity costs for their time, they invest more in constructing bunds and terraces. On the other hand, Bunch (2002) suggests that successful soil management programmes in Central America were responsible for keeping people, who otherwise would have migrated, on farms. There is also the possibility that labour-intensive LEIT might reduce seasonal or permanent migration by offering local employment on other people's farms. This seems to be happening to some extent in Niger with the use of planting pits (Hassane et al, 2000).

The Niger case highlights the fact that many of the labour requirements of LEIT may be met by hired labour. Such opportunities are more likely to be available to households that have cash to hire labour, derived either from their farming operations or from off-farm activities. Winters et al (2004) found that those households in Ecuador with higher off-farm income in Ecuador were more willing to invest in resource-conserving technology such as terracing (although they were less likely to participate in the conservation project). Farmers in southern Mexico who had significant commitments to off-farm labour were less willing to experiment with cover crops (Soule, 1997). Brown and Korte (1997) found that an area in the Philippines with high uptake of hedgerow terracing was close enough to a city for family members to be able to earn off-farm income and still maintain a place in the countryside, encouraging investment in the farm. The study by Tiffen et al (1994) in Machakos, Kenya, showed that the availability of migratory (and local) wage labour opportunities provided income that was invested in farming and in significant soil conservation activity. The success of this model in other cases would depend upon whether off-farm income was sufficient to invest in farming, the degree to which direct supervision was required, and whether there were adequate incentives for investing in farming.

Off-farm labour opportunities may have a negative impact on the utilization of LEIT by the poorest households because they draw labour from the farm and because they affect the possibilities for learning new agricultural techniques. Crowley and Carter (2000) describe the downward spiral of unskilled rural migrants from western Kenya who have too little time or cash to invest in improving their own farms and end up working for others or migrating in search of employment. Even when farmers are not forced to look for off-farm work, the availability of non-farm opportunities may be a disincentive to farm investment. This is especially the case if the technology involves organizing extra labour or substituting labour and time for inputs that can be purchased with off-farm earnings. Boyd and Slaymaker (2000) discuss several African cases where off-farm employment opportunities were an important explanation for farmers' lack of interest in soil and water conservation. Neill and Lee (2001) found that farmers with more non-farm income were less likely to maintain the use of cover crops (although the reasons are complex). In Madagascar, those farmers who tried SRI techniques and then abandoned them were more likely to be those with higher dependence upon off-farm income (Moser and Barrett, 2003). Rice farmers living near the main road in an Indonesian village, and who had other income sources, were much less interested in learning the techniques of IPM (Winarto, 2002).

The relationship between rural livelihood strategies and the propensity to use LEIT is very complex. In a study in Tanzania, Birch-Thomsen et al (2001) found no discernible relationship between soil conservation strategies and livelihood strategies. As Scoones (2001) shows, the allocation of household assets among various activities, including the use of resource-conserving technology, depends upon many different factors and varies not only between farmers but also between fields. Similarly, Crowley and Carter (2000) describe niches of investment in soil conservation depending upon field conditions, crops and household resources.

Labour: Summary

In summary, labour is one of the most important factors determining the breadth and equity of LEIT adoption; but it is essential to identify the nature of the labour component of any LEIT innovation. Some examples of LEIT require relatively little extra labour and some actually reduce the labour requirements of farm management. But other examples require significant labour investments to establish the technology and/or to maintain it. The timing and nature of that labour are crucial to understanding the prospects for success. In addition, LEIT often requires time to learn new skills. The importance of labour and time for managing LEIT helps to explain why the poverty impact of this type of technology may be less than expected. Poor households do not necessarily have surplus labour or time to invest in LEIT, nor do they have the financial resources to hire labour. For those households that are able to maintain farm operations while at the same time generating off-farm income, the situation may be different. In some cases the extra cash can be invested in more labour for LEIT, while at other times the existence of non-farm employment lowers the incentives for investing in new farming skills. These incentives for farming, and for learning about LEIT, are related to the human capital assets of the farm, which are examined in the following section.

Human capital

The capacity to learn techniques associated with LEIT may be associated with locally acquired skills or innovative attitudes, formal schooling or the external provision of specific learning opportunities. The utilization of LEIT may also depend upon or, in turn, foster an understanding of the environmental consequences of agricultural technology.

Education and skills

There is limited information about the relationship between education and the adoption of LEIT in developing countries and no clear conclusions can be drawn. Education had no effect on farmers' initial participation in a soil conservation project in Ecuador, but was strongly related to the intensity of adoption of techniques promoted by the project (Winters et al, 2004). In Madagascar, rice farmers who adopt (or who try and then abandon) SRI tend to have somewhat more education than non-adopters (Moser and Barrett, 2003). In the case of indigenous soil conservation techniques, Pender and Kerr (1998) found a significant relationship between education and investment in two villages in India. On the other hand, there is no relationship between the adoption of a velvet bean cover crop and education in Benin (Manyong et al, 1999). In a Kenyan study, the uptake of agroforestry techniques involving biomass transfer was not related to farmers' education, while the adoption of improved fallows was associated with higher levels of schooling (Place et al, 2003). In those cases where there is a positive correlation between education and adoption, we can only speculate on whether this is due to the formal skill requirements of the technology, an appreciation of the rationale for the resource-conserving technique or simply a random association.

Studies from industrialized countries exhibit a somewhat more consistent association between formal education and adoption of LEIT. Traoré et al (1998) found education level (but not experience or farm size) to be associated with the adoption of soil and water conservation practices in Quebec. Pingali and Carlson (1985) found that farmer education and experience (age) had the greatest impact on the abilities of North Carolina apple farmers to take advantage of IPM; attendance at formal courses was also significant. Similarly,

California pear growers with higher levels of education were more likely to adopt the various components of an IPM programme (Ridgley and Brush, 1992). On the other hand, adopters of an IPM programme for tomatoes in California had no more education than non-adopters (Grieshop et al, 1988). In the US, those farmers who are aware of the techniques of precision agriculture (including grid soil sampling, yield mapping and variable rates of input application) tend to have more education than those who are not aware, although within the former group level of education has no effect on adoption (Daberkow and McBride, 2003).

One of the advantages often posited for LEIT is its capacity to stimulate further innovation. There is, however, relatively little evidence in the literature that demonstrates how the experience of utilizing particular instances of LEIT leads to further innovation. Fujisaka (1992) examines how farmers in the Philippines adapted hedgerow technology by experimenting with different species and management techniques. Much discussion on this subject tends to be anecdotal. A recent collection of essays offers examples of farmer innovation in Africa in areas such as soil management that are largely unrelated to any contact with formal research or extension (Reij and Waters-Bayer, 2001). Scoones (2001) also discusses farmer innovation, particularly in relation to soil management, but tends to put more emphasis on links with extension and research (while pointing out that much conventional research and extension are currently not up to the task of contributing to this type of interchange). Farmers in Guatemala who participated in a LEIT project often applied the new techniques (such as conservation tillage) to small fields of low potential rather than to their main subsistence plots. There is little evidence that these farmers later expanded their experiments (Vaessen and de Groot, 2004).

External support

Several observers point to the need for long-term extension advice in the conduct of LEIT in order to encourage and support farmer experimentation and innovation. Commenting on the evolution of upland farming systems in the Philippines, which implies the need to continually develop new strategies, Cramb et al (2000) suggest that conservation technologies require a continued extension presence. A study of IPM in the US concludes that 'farmer uncertainty is severe enough in many cases that considerable, prolonged contact with extension agents may be necessary' (Cowan and Gunby, 1996, p539). Heong and Escalada (1999) discuss the importance of changing farmers' beliefs about pest damage and their tendency to overestimate yield loss in the introduction of IPM in Vietnam. Orr (2003) describes the limitations of farmer experimental capacities for supporting the uptake of IPM in Africa and suggests that (formal) knowledge of insect pests and their habits is a priority. Neill and Lee (2001) suggest that the diffusion of resource-conserving technology in Honduras will be sporadic until farmers have a better understanding of system dynamics and the specific details of nutrient cycles, nitrogen fixation and herbicide effects.

There are several examples where it is clear that LEIT requires skilled and sustained technical input from extension or other service providers. Place et al (2002a) found that agroforestry-based technology on soil fertility was used more extensively in villages in Kenya and Zambia that had served as pilot projects and thus had access to considerable technical backstopping. A study of the adoption of SRI in Madagascar found that most farmers learned the techniques from extension agents, which raises questions about the extension intensity of this technology (Moser and Barrett, 2003). A number of NGOs that originally promoted conservation farming in Zambia have now retreated, in part due to a shortage among their staff of the management and agronomic skills required for promoting the technology (Haggblade and Tembo, 2003).

Understanding the environmental consequences

The adoption of LEIT is not only more likely when farmers have access to the requisite skills, provided by formal training or informal learning, but in many cases farmers need to be more cognisant of long-term effects and returns, and to appreciate the environmental and conservation motives of the technology (a discussion of informal learning is provided below in the discussion of social capital).

It is inaccurate to characterize all LEIT as providing benefits only in the long term; but certainly one of the greatest challenges for technologies such as soil conservation is to provide opportunities for farmers to incorporate the value of future returns in their decision-making. This involves both understanding how farmers see returns on investments and how they envision farming more generally. Farmers' willingness to make investments in labour or other inputs for technology whose pay-off may be realized only several years into the future depends, in part, upon their perception of the returns. A number of studies have been done on farmer discount rates, and these will not be reviewed here in any detail. One study in India (Pender, 1996) found that farmers' discount rates (as measured in an experiment) significantly exceeded local market interest rates, although there was considerable variability in the responses. The high discount rates implied 'a rational disregard by most households in the study villages for all but very near term impacts of decisions' (Pender, 1996, p290). Lumley (1997) estimated discount rates in the Philippines and found them below the (exceptionally high) local interest rates. Surprisingly, the poorer households tended to have lower discount rates; but there was no consistent relationship between a household's perceived discount rate and its adoption of soil conservation practices.

The tendency for the poor to base investment decisions on high discount rates raises serious challenges for introducing some types of resource-conserving technology. Where success is noted, it is often in those components that provide more immediate benefits. Murray (1994) describes how farmers in the Dominican Republic valued soil conservation measures for the short-term impact of lessening the rilling and gullying that affected their fields, rather than for any long-term productive pay-offs. Versteeg et al (1998) describe how farmers in southern Benin adopted a cover crop of velvet bean, originally introduced for long-term soil fertility maintenance, because it helped them with the immediate problem of controlling speargrass, a difficult weed. In the north of the country, where there are fewer weed problems and more land is available, the adoption of the cover crop is significantly lower (Manyong et al, 1999).

Beyond estimating formal discount rates, it is also important to understand farmers' own experiences and perceptions about resource management. Leach and Mearns (1996), for instance, challenge the view that Africa is subject to massive natural resource degradation, much of it caused by the environmentally uninformed practices of farmers and herders. They propose that local practice, especially when viewed over the long term, is better at conserving natural resources than many of the practices induced by policies imposed by governments and donors. On the other hand, Baland and Platteau (1996) argue that indigenous communities are not inherently conservationist and point to the necessity of changing traditional beliefs:

> Awareness of ecological stress under conditions of increasing and continuous human pressure on the environment may grow even if only slowly and even if it typically requires visible signs of depletion/degradation to be stimulated. This awareness is therefore likely to develop more rapidly in those societies in which 'a sense of limits' has already pervaded people's minds due to previous experiences of scarcity. Moreover, it emerges more easily with respect to localized, visible and predictable resources than with respect to resources showing the opposite characteristics (Baland and Platteau, 1996, p233).

In addition to the importance of estimating the long-term effects of some LEIT, an appreciation of environmental challenges, and the contributions of LEIT, can influence the spread of the technology. An understanding of threats to the environment has sometimes been found associated with the adoption of LEIT in industrialized countries. Traoré et al (1998) found that Quebec farmers' conservation practices were influenced by the degree to which they perceived environmental degradation as a problem. Similarly, Burton et al (1999) found that organic farmers in the UK tended to be more concerned about the environment and the sustainability of the food system. On the other hand, Lockeretz and Wernick (1980) found that Midwest US organic growers were more likely to be motivated by a specific issue (such as their health or a particular soil problem) rather than by philosophical, religious or ideological concerns. It may take much more than generalized concerns about the environment to encourage the use of resource-conserving technology. An examination of tree-planting practices on farms in Australia found that:

> ... environmental orientation or concern about an environmental problem had no observable effect on the decision to plant trees ... pro-environmental attitudes will not translate into pro-environmental behaviour unless there are economic or other benefits associated with the behaviour (Cary and Wilkinson, 1997, p19).

Even if pro-environmental attitudes are important, these are slow to emerge in developing countries. In examining agricultural development in Kerala, Veron (1999, p180) found that 'environmental degradation and agricultural stagnation do not seem so much the problems of farmers as of natural scientists, economists and a few activists ... the normative concept of sustainable development has not become a general cultural value in Kerala ... despite the high levels of education and social awareness.' In a study of a project promoting the use of bio-pesticides in Andhra Pradesh, Tripp and Ali (2001) found that, while farmers were willing to experiment with alternative pest control measures, they were not motivated by the same concerns as the environmentally minded NGOs that organized the projects.

Even when farmers accept the environmental rationale for one set of practices, this does not guarantee the development of a broader sustainability perspective. Bunch (2002) describes how low-input soil fertility innovation in some areas of Honduras has been accompanied by deforestation and high pesticide use for horticultural crops. Holt-Giménez (2001) discusses how Central American farmers are drawn to organic and fair trade markets by economic rewards, even when these techniques may be associated with monoculture and deforestation.

Human capital: Summary

There are several significant relationships between human capital and the uptake of LEIT. Much of LEIT implies learning new skills, and in the relatively few cases where the question has been asked, farmers' level of education is sometimes associated with the use of LEIT. This relationship is also commonly found in studies of the adoption of conventional technology; but in neither case is it usual to see a clear explanation as to why the ability to take advantage of new technology might be dependent upon a certain level of education. The relationship underlines the fact, that although LEIT is based on the use of local resources, it often depends upon technical expertise and advice. Some observers make the argument that LEIT requires an even better-prepared extension service than conventional technology. It is a challenge to ensure that the poor have access to this type of advice and the time and resources to take advantage of it. However, given the wide range of examples of both LEIT and conventional technology, it is unwise to suggest that LEIT can be characterized as particularly knowledge-intensive. An additional challenge for the promotion of some LEIT is to ensure that farmers base their decisions on factors beyond the calculation of short-term economic gains. However, experience to date gives few examples of farmers using lower discount rates, or taking account of environmental concerns, in their decisions to use LEIT. Not surprisingly, it is difficult to promote these attitudes in the context of promoting particular technologies. One way in which wider support for more environmentally sustainable agriculture may be developed is through political movements, the subject of the following section.

Political movements in support of LEIT

Some LEIT requires at least some farmer organization, as well as a change in attitude about the nature of agriculture. A potential role for political movements in the promotion of LEIT should therefore not be unexpected or discouraged. Indeed, effective and representative political organization within farming communities may be more important for the future of farming than the acceptance of any particular technology (Chapter 3 discussed the blind spot regarding political organization in most discussions of social capital and agricultural change). The degree to which a resource-conserving technology is related to a wider political arena will also have an influence on how farmers view its potential returns and value.

One example of the close correlation of LEIT promotion with political action is the spread of soil and water conservation techniques (particularly stone bunds) in Burkina Faso (Smale and Ruttan, 1997). A charismatic leader built upon Mossi traditional temporary male work groups to develop a nation-wide movement (*Groupements Naam*) that now performs a range of community and income-generation services. The emergence of the movement was linked to a period of severe drought when various community action activities and much NGO investment were in evidence. The resurrection and improvement of indigenous soil and water conservation structures became an important part of the movement. The movement is now well developed, receiving local and international support, and co-exists (and at times competes) with other NGO and government efforts at local group formation (Atampugre, 1993). The conservation measures that are established under the aegis of *Groupements Naam* and other state and NGO strategies make an important contribution to improving and stabilizing cereal yields; but it would be difficult to distinguish strict cost-benefit results from political considerations in understanding farmers' motivations for this type of investment.

Dilts (2001, p19) discusses various activities in Indonesia that followed the introduction of IPM, reporting that 'an array of IPM farmer institutions has sprung up across the country'. To what extent this is a spontaneous evolution or, rather, the product of government (or donor) support for IPM is not clear; but Dilts mentions farmer congresses and associations with links to local government. One case even describes a local religious leader who incorporates the principles of IPM within his lessons. To the extent that the movement is spontaneous, it is fair to ask the degree to which it profits from the human and social capital contributions of farmer field schools or, on the other hand, simply links a move towards the decentralization of government services with donor funds for local-level IPM activities (Fakih et al, 2003).

Various types of technology (including LEIT) may be promoted by national governments (often in concert with donors). Such promotion (and association with the regime in power) may have contradictory effects on technology adoption. In many cases, it is undoubtedly useful to have the forces of the government behind a new technology. Keeley and Scoones (2000) analyse the recent history of Ethiopian government policy for technology promotion. Much of this has been dominated by a Green Revolution philosophy, although in a variety of manifestations (e.g. Swedish rural development efforts under Haile Selassie, the state farms of the socialist Derg regime, and the current government's support for the seed and fertilizer technology packages of Sasakawa Global 2000). During recent years, there has also been an environmental interest in Ethiopian agricultural policy, but manifested mostly through an imperfectly documented environmental crisis and the promotion of physical soil conservation measures. These conservation efforts tend to rise and fall related to the presence of donor food aid programmes; but the environmental issues remain an important element of government policy. The authors go on to describe a third emerging force in Ethiopian agricultural policy: a participatory agricultural and natural resource management strategy (conducive to LEIT). Whether this represents a triumph of grassroots common sense over top-down strategies, or is merely the latest manifestation of donor and government fashions (this one being fuelled, in part, by a tendency of donors to channel assistance through NGOs and an attraction for decentralization) remains to be seen.

In some cases, the promotion of a technology by an unpopular regime may have a negative effect. Perhaps the cases most relevant to LEIT are the numerous historical examples from Africa where colonial administrations attempted to impose soil conservation measures and found them widely opposed or even sabotaged. As early as the 1920s, regulations on natural resource use were introduced to the Mbeere who live east of Mount Kenya. These prohibited planting on hillsides or burning grass. Later, the colonial administration's compulsory construction of terraces was accompanied by arbitrary and sometimes corrupt enforcement of regulations (Little and Brokensa, 1987). In Kenya during the 1940s, the Akamba, suspicious that government efforts to promote soil conservation were part of a strategy to annex native land, refused to allow work parties to construct terraces on their farms and threw themselves in front of government tractors sent to aid in terrace construction (Tiffen et al, 1994).

On the other hand, the association of unpopular regimes with the Green Revolution has sometimes been an important factor in orienting the opposition towards LEIT. An outstanding example is the network of NGOs that arose in the Philippines in opposition to the Marcos regime, especially after the imposition of martial law in 1972. The regime was tightly linked to the promotion of high-input Green Revolution technology, and it is not surprising that the opposition, often working with farmers in traditional and marginal production zones, focused on environmental issues and technologies promoting self-sufficiency (Miclat-Teves and Lewis, 1993). This political element, featuring the promotion of technologies in opposition to the regime in power, can be an important factor in determining farmers' attitudes to LEIT. For instance, it can be argued that the promotion of IPM in Indonesia built on resistance to the extraordinary control exercised by the government extension service over rice farmers. IPM not only represented an opportunity to decrease dependence upon purchased chemicals, but also to exercise more individual control over the farm management practices mandated by government package programmes (Fakih et al, 2003).

When a particular type of technology achieves some measure of political symbolism, difficult choices may have to be made and farmer pragmatism may be subjected to other pressures. Branford and Rocha (2002) describe the history of the Landless Movement in Brazil, in which groups of poor people have gained access to former commercial farming property through occupation or land reform. There was an initial tendency for these groups to attempt to beat the commercial farmers at their own game and adopt high-input strategies. A combination of management disappointments and the availability of political support from a 'green' constituency has shifted many groups towards low-input or even organic production.

Bebbington (1992) describes the choices open to various peasant political groups in Ecuador. Some resent the attempt by outside supporters to impose 'traditional' agricultural methods when they aspire to enter mainstream agricultural production, while others recognize the value of such technology in affirming and validating an indigenous identity. The decision facing farmers is a difficult one. As Bebbington et al (1993, p183) ask: 'what ... would happen if the socio-political logic of a popular organization suggests a resource management practice that differs from the ecological and economic ideal'?

Social capital and information flow

We also need to consider how the flow of information and advice for LEIT can be made most effective. This section examines the experience of LEIT efforts related to social capital, including farmer-to-farmer information transfer, networks, building on existing groups, and the creation of new groups.

The movement of information

There is much evidence that information about new technology travels from farmer to farmer, and it is important to understand the extent to which this is the case for LEIT, as well. However, there may be challenges for those types of LEIT that are particularly information-intensive. Cowan and Gunby (1996) discuss 'network externalities' (the utilization of information from neighbours) and cite evidence that farmers consider their neighbours' pest control practices when formulating their own strategies. However, they are concerned that the complexity of IPM limits the possibilities for such transfer in the US. Cramb et al (2000) discuss the possibilities and limitations for farmer-to-farmer extension of conservation technology in the Philippines. The most successful project examined in that study was found in a fairly large community where farmers easily formed working groups, led by farmer instructors, to institute soil conservation measures. The other less successful sites were in communities composed of isolated hamlets, or where there had been internal rifts that limited farmer-to-farmer communication.

Vaessen and de Groot (2004) found no evidence that LEIT innovations, such as conservation tillage, introduced to Guatemalan farmers spread to neighbours despite the fact that the project participants described their experiences to other farmers. There is also no indication that West African farmers who adopted alley cropping transferred the technique to other farmers; but the diffusion of cover crops for weed control (an easier technology to understand) was significant (Douthwaite et al, 2002).

There is considerable hope that the farmer field school (FFS) approach used for IPM will either stimulate farmer-to-farmer transfer or help to develop local farmer extensionists. There is some preliminary evidence for this. Kimani et al (2000) describe the experience of a pilot project in FFS for pest and disease management in vegetables in Kenya, which placed particular emphasis on the design and conduct of simple experiments to test possible interventions. A follow-up assessment with two groups 18 months later found that each FFS farmer had shared some ideas with an average of four other farmers and that these farmers had usually implemented one or more of the new ideas. In a study in the Philippines, farmers in a village voluntarily placed themselves in one of three groups: an FFS that spent a season learning techniques of IPM; a group that received advice on reducing insecticide use; and non-participants. Farmers in the FFS learned the most; but there was evidence of transfer of knowledge between farmers in different groups (Price, 2001). K. Jones (2002) gives preliminary evidence from Sri Lanka of the transfer of knowledge from farmers who participated in FFS to others in neighbouring villages.

On the other hand, Rola et al (2002) found that the extent of farmer-to-farmer transfer of IPM methods and principles in the Philippines was quite low and that the strategy of developing farmer trainers did not result in the expected degree of information transfer. The study included a 'snowballing' technique in which researchers sought out the farmers to whom FFS graduates had reported passing information. The study concludes that informal interactions between FFS graduates and others are not sufficient to ensure diffusion of IPM principles, perhaps because they are more abstract than the type of farming information usually passed among farmers. Quizon et al (2001) discuss the Philippines case as well as studies from Indonesia that indicate that the strategy of placing greater reliance upon farmer trainers for IPM is not working as expected.

Networks and groups

The effectiveness of farmer-to-farmer information flow for LEIT is related to the strength of local networks. Ridgley and Brush (1992, p373) found that IPM adoption in California was associated with 'families who had lived and farmed in the same area for several generations [and] have dense social relationships established through farming connections and social events'. Burton et al (1999) describe informal networks of organic growers in the UK who consult each other. Rola et al.(2002) found that in those instances where FFS graduates in the Philippines shared IPM knowledge with others, over one third of the instances were with relatives and more than half were with close neighbours.

With respect to formal organizations, it is common to read that success in promoting LEIT often profits from previous 'social capital', the existence of effective village organizations or previous successful community activities.

Winters et al (1998) found that those farmers who were active participants in conservation project activities in Ecuador tended to exhibit a higher level of group membership; indeed, the NGO in charge of the project used superior community organization as one of its criteria for site selection. Sanders (2000) found that villages in China that adopted 'ecological agriculture' were those with superior organization and strong leadership. Belknap and Saupe (1988) found that membership in community organizations was related to the adoption of conservation tillage in Wisconsin; the authors assume that this is related to expanded opportunities for learning about the technology. Noordin et al (2001) comment on the fact that most villages in Kenya have various groups and that building on existing groups, rather than forming new ones, often makes sense for promoting agroforestry activities. Brown and Korte (1997) show how traditional labour exchange groups were utilized to hasten the uptake of hedgerow terracing in the Philippines. Many more adopters than non-adopters of SRI in Madagascar belong to a farmer association (Moser and Barrett, 2003).

Despite the recognized importance of building on extant community organization, there are many instances where the promotion of LEIT includes forming new groups or structures. The absence of any significant farmer organizations in communities in Zanzibar required the formation of special groups for IPM. Some of these groups collapsed because of various problems; but the majority proved to be viable for the purposes of the project (Bruin and Meerman, 2001). Cowan and Gunby (1996) discuss the necessity of coordinating the adoption of IPM practices in the US; single adopters will probably not see benefits until their neighbours also reduce pesticide use. The rapid spread of conservation tillage in Brazil was due, in part, to the formation of Friends of the Land clubs that facilitated farmer-to-farmer exchange of experience, provided support to farmers experimenting with the technology and allowed access to outside expertise. Many of these clubs were assisted by support from the chemical companies that were selling the herbicides used for conservation tillage. However, once farmers became familiar with the techniques, interest in the clubs tended to wane unless other activities were included (Landers, 2001).

It is important to consider the viability of farmer groups formed specifically for the purpose of introducing and supporting the use of LEIT. Farmer field schools have been utilized for IPM, with the presumption that they might continue to function as groups after the initial project period. Owens and Simpson (2002) find evidence for this in West Africa in cases where the FFS was the only significant local organization, but not when other organizations already existed. Winarto (2002) describes how a NGO in Indonesia that successfully introduced IPM to rice farmers helped to form an alliance of farmers' associations that campaigned to remove pesticides from the government credit package and lobbied for lower prices and more timely delivery of fertilizer. However, there is little evidence that the majority of FFS formed in Indonesia for introducing IPM survive beyond the initial season of training. In some cases, there are hopes that a group developed for LEIT may take on broader responsibilities; but again there is only anecdotal evidence. Brown and

Korte (1997) discuss the problems in the expected transition in the Philippines from an externally financed project that successfully elicits local organization for specific soil conservation activities to a more generalized people's organization with diffuse goals and fewer external connections.

Despite considerable interest in the relationship between social capital and LEIT, the literature provides relatively little evidence on this subject, and little reason to believe that the social capital of the poor is necessarily sufficient for the efficient diffusion of new technology. There is some transmission of information about LEIT from farmer to farmer; but existing networks are not necessarily extensive enough, and the information to be transmitted is sufficiently complex for there to be little documentation, to date, of widespread, spontaneous diffusion of LEIT. The few cases where local transmission of information occurs most effectively seem to be those where an intensive community-level project is in operation, rather than a broader extension strategy. The existence of formal organizations through which LEIT promotion can be channelled is an obvious advantage, although such conditions are more likely for the more affluent farmers. The hopes that groups formed specifically for introducing LEIT, such as farmer field schools, will assume a life of their own and serve ever broader purposes have yet to be realized.

Summary

This chapter has reviewed literature related to the uptake of various types of LEIT. The literature on this subject is not extensive; hence, any conclusions must be seen as tentative. The chapter concentrated on examining the nature of LEIT adoption and asking if the patterns observed correspond with the aspirations of LEIT to address the needs of resource-poor farmers who are overlooked by many other efforts at agricultural technology generation.

Although much of LEIT is aimed at marginal farming areas, neither the success rates observed in those areas nor the adoption patterns indicate that LEIT has been particularly effective at reaching the poor. Even in areas with significant success, the preponderance of evidence indicates that LEIT does not escape many of the biases typical of other agricultural technologies. It is often relatively wealthier households (albeit frequently in marginal farming areas) which are able to take advantage of the technology. Some of these patterns may simply be due to the fact that the wealthy are often the first to adopt and others may soon catch up. In addition, some of the problem may lie with the way in which LEIT efforts are targeted or organized. But a significant part of the explanation surely lies with factors such as economic incentives, and access to labour, skills and contacts that favour better-resourced households.

The literature shows that LEIT is more successful in areas and among farmers who participate in viable agricultural markets. Not all LEIT requires extra labour; but when it does, it is often wealthier farmers who are able to provide it, either from within the household or by hiring labour. The labour investment in LEIT is further complicated by household strategies favouring off-farm employment. This may provide additional cash for investment in the farm, on the one hand, but may reduce the time or incentives for learning the skills of LEIT, on the other. LEIT may also demand time for learning new skills, and again the poorer households may not have these capacities.

The skills required for LEIT help to explain why adoption is occasionally related to educational status. The skills of LEIT are unlikely to be found in school curricula, however. Farmers can gain these skills if they have the time and opportunity to experiment, and if they have access to good-quality technical advice. An understanding of the long-term and environmental advantages of LEIT also helps to provide incentives for adoption. Again, these conditions are less likely to be found among poorer households. At times, useful support for LEIT is found in the ideologies of political movements that seek to represent the interests of more marginalized members of the farming community.

Some LEIT has been shown to spread from farmer to farmer; but there are questions about the degree to which much of this information is easily transmitted through traditional informal channels. The introduction of LEIT is often accompanied by group action, using either extant organizations or forming new ones. These are undoubtedly helpful; but there are questions about the viability and efficiency of many of these groups.

The biases and bottlenecks observed in the diffusion of LEIT should not be used to reduce support for the efforts to develop and introduce this kind of technology. Nevertheless, they raise serious questions about whether LEIT should be seen as a separate class of technology, associated with a unique ethos and vision of farming. Detailed case study analysis that contributes to addressing some of these questions is presented in the next three chapters.

5

Learning from Success: Revisiting Experiences of LEIT Adoption by Hillside Farmers in Central Honduras[1]

Michael Richards and Laura Suazo

Introduction

Background

Various stories of the successful adoption of low external-input technology (LEIT) by small farmers are reported in the literature on sustainable agriculture. But these success stories are often anecdotal, focusing on individual 'star farmer' case studies. Here we attempt a more systematic analysis of one of these stories – the promotion of hillside technologies for small farmers in central Honduras by World Neighbors and associated non-governmental organizations (NGOs) during the 1980s and 1990s.

It is estimated that hillsides cover about 85 per cent of Honduras, and that at least 30 per cent of the population live in these areas, 90 per cent of them below the poverty line (Jansen et al, 2003). The highly skewed land distribution, the degrading hillside soils, the rapidly expanding population and the diminishing agricultural frontier have made the traditional slash-and-burn farming system unsustainable. Most smallholder farmers in central Honduras now rotate their annual crops on the same land, growing maize (and sometimes beans) in the first growing season (*primera*), and beans (and sometimes maize) in the second rains (*postrera*). Farmers with irrigation (mainly using aspersion systems) and access to markets tend to grow vegetables; most hillside farmers grow some coffee; but coffee prices have slumped since 1997, increasing the pressures on basic grains and vegetable farming.

Continuous cropping of hillsides is associated with a downward spiral of soil erosion, declining yields, increasing costs and, ultimately, abandonment, so that many farmers become farm labourers or migrants. Apart from the poverty impacts, this process has major environmental implications, including increased vulnerability to extreme weather events such as Hurricane Mitch (Walker and Medina, 2000). These problems have been exacerbated by the macroeconomic context and policies favouring agricultural modernization in the valley areas (Falck and Díaz, 1999).

Responding to this crisis, many NGO and some government programmes over the last two decades have promoted sustainable hillside farming through a range of methods, enhancing soil and water conservation and fertility. During the early 1980s, a consortium of two NGOs, World Neighbors (WN) and the Co-ordinating Agency for Development (ACORDE), and the Ministry of Natural Resources started to experiment with and promote LEIT practices for hillside farming, some of them 'imported' from a WN project in Guatemala and others developed by ACORDE. In the 1990s, these efforts were consolidated by the Association of Sustainable Agriculture Advisers (COSECHA), a NGO founded by former WN members. The approach developed by these NGOs has been widely adopted (or adapted) nationally and regionally over the last 20 years.

Objectives and structure of the study

The main objective of the study is an empirical analysis of enduring impacts and determining factors in the processes of LEIT adoption and dissemination, particularly the roles of human and social capital. It tries to assess whether pre-existing human and social capital contributed to LEIT adoption, and if the LEIT experiences contributed, in turn, to the development of social capital so that a mutually reinforcing cycle of technological and social progress was established. An expression of this idea is that a successful LEIT project can 'create in villagers a strong motivation to innovate and work for the common good' (Bunch, 1999).

The chapter is structured as follows. This section introduces the projects and technologies, followed by a description of the research methodology. The next section presents some general results of the household survey, including the enduring levels of technology adoption; this is followed by an analysis of the factors determining adoption. Subsequent sections consider the issues of farmer innovation capacity as an aspect of sustainability; information flow and technology dissemination; project impacts on social and human capital; and the use of external inputs by LEIT adopters. The final section presents some conclusions.

Introduction to the projects and research area

Two main project experiences were selected for study, one from the 1980s and one from the 1990s, in contiguous municipalities of the Department of Francisco Morazán (see Figure 5.1). The 1980s project was the Cantarranas Integrated Agricultural Development Programme implemented by WN and ACORDE, popularly referred to as the ACORDE project. This was implemented in 35 communities in Cantarranas Municipality between 1987 and 1993. The second project was the Guaimaca Integrated Agricultural Development Programme implemented by COSECHA in 30 communities in Guaimaca Municipality between 1992 and 1999. These projects show visible and enduring impacts (in terms of current management practices) and have influenced many important national and regional agricultural and rural development projects – for example, the United Nations Food and Agriculture Organization (FAO)-supported Lempira Sur project in Honduras.

Figure 5.1 Study communities in Franciso Morazán Department, Central Honduras

Project design was based on five main principles (Bunch, 1999): motivating and teaching farmers to experiment with new technologies, initially on a small scale; using rapid and recognizable success rather than incentives or subsidies; focusing on technologies using local resources; concentrating on a small number of technologies; and training village leaders as voluntary or paid extensionists. Farmers were taught a 'minimum' of basic soil science, encouraged to share results with others and motivated to do this with minimum outside support. The training emphasized individual farmer experimentation or innovation, including the use of simple control plots. Extension methods included village meetings and talks, individual farmer visits, group demonstration plots and visits to outstanding farmers or communities. The vast majority of participants were male; but women's groups were also formed and trained in food preparation, basic healthcare and the development of 'family orchards'. The projects normally lasted four to five years.

The project area is typical of much of central Honduras – relatively homogeneous in topography and climate, but with substantial variation in population density, access to markets and agricultural practices (Pender and Scherr, 2002). The climate is sub-humid tropical with annual rainfall in the range of 1000–2000mm. More than 90 per cent of the region is on hillsides; soils are generally thin and of poor quality, with serious levels of soil erosion, estimated at between 22 and 46 tonnes per hectare (Hellin and Schrader,

2003). The altitude ranges from 300m above sea level to 1500m; but the study focused mainly on communities in the range of 1000m–1200m.

The project technologies

The main project LEITs were in-row tillage (IRT), locally known as 'minimum tillage', the use of green manure cover crops (GRMs), live barriers or contour hedgerows, and organic fertilizers (including chicken manure, composting, coffee pulp, sugar cane bagasse and ash). Technologies promoted on a lesser scale included agroforestry practices and physical soil and water conservation practices such as stone walls and drainage ditches. Given the range of technologies and the limited research budget, we decided to focus on the technologies considered to be of prime interest in these projects: IRT and GRMs, and, to a lesser extent, live barriers, noting that the use of organic fertilizers was closely related to IRT.

The main characteristics of IRT are manual in-row cultivation of the soil within a radius of about 25cm from each plant, leaving the area between rows uncultivated; contour cultivation, usually after using an A-frame; use of residual organic matter from previous crops and weeds; and the gradual (over several years) or one-off development of mini-terraces (Arrellanes, 1994; Hesse-Rodríguez, 1994; Bunch, 1999). The main benefits of IRT are recuperation of soil fertility without chemical fertilizers; concentration of organic matter in the crop root zone; weed control; reduced erosion; and conservation of soil humidity. GRMs refer to leguminous plants used to improve soil fertility by the production of organic matter and to suppress weeds. The predominant GRM was velvet bean (*Mucuna pruriens*); but some farmers also used *Carnivalia ensiformis*, *Dolichos lablab* and *Cajanus cajan* (*Gandúl*). Live barriers refer to contour-planted strips, mainly of biannual plants or grasses, for the retention of soil and moisture. The most commonly planted live barrier plants were 'King grass', valeriana, lemon grass and vetivier grass, as well as crops (sometimes interplanted) such as sugar cane and pineapple.

Methodology

Selection of study communities

Following discussions with project extensionists and visits to a range of the project communities, we selected communities in the mid-high altitude range (1000m–1400m) where there was reasonable success at the time of the projects. We studied five communities in the ACORDE project area and three in the COSECHA project area, two of which were control communities. We classified the communities into three regions for the purpose of the study: region 1 or San Juan de Flores (SJ de Flores), composed of communities 1 to 4 in the ACORDE project area; region 2 or Guaimaca, composed of communities 6 to 8 in the COSECHA project area; and region 3, the higher altitude (1400m) ACORDE community of Guacamayas. Selection of the control communities in the ACORDE project area was difficult because of the blanket coverage of the project in Cantarranas Municipality. There was no alternative to choosing as a control the outlying hamlets of a project community (Joyas Del Carballo). These hamlets, which we have called Joyas Del Carballo no 2, have mainly developed since the ACORDE project, so that it is possible to argue that they were hardly affected by the project. But at the same time, three farmers from Joyas Del Carballo no 2 proved to have been project participants, so it is clearly a far from perfect control community. In the case of region 3 (Guacamayas), we were unable to identify a control community.

In most of the analysis, regions 1 and 2 are combined because of the similarity of their agro-ecological conditions and development constraints. Region 3, Guacamayas, is usually examined separately in this study because of its rather unique characteristics. Most of the communities were in the buffer zones of two protected areas surrounding cloud forests, La Tigra National Park and El Chile Biological Reserve.

Table 5.1 *Communities studied and sample size*

Community name	Region	Project/ control	Project dates	Approximate number of farmers[a]	Sampled farmers
1 Joyas Del Carballo no 1	1	ACORDE	1983–1987	33	19
2 El Hato	1	ACORDE	1989–1991	23	13
3 El Carbón	1	ACORDE	1987–1991	51	29
4 Joyas Del Carballo no 2	1	Control for 1–3	–	36	17
5 Guacamayas	3	ACORDE	1986–1991	37	22
6 San José	2	COSECHA	1994–1999	61	31
7 La Aserradera	2	COSECHA	1992–1999	70	30
8 El Zapote	2	Control for 6–7	–	70	17
Total	–	–		381	178

Note: [a]Participant and non-participant farmers still resident in the communities.

Selection of sample farmers

A rule-of-thumb approach was to select 50 to 60 per cent of the participant and non-participant farmers in each community still living in the communities. (Since the projects finished 5 to 15 years before this study, we found that many of the participants and non-participants had left. In several of the communities we therefore had a relatively small sample frame.) Participants were defined as including all who attended at least a few meetings or training sessions. There was, therefore, a wide range of participation: 52 per cent of the 79 participants undertook trials and stayed to the end of the projects, 14 per cent undertook trials but then dropped out; 4 per cent received training but did not undertake trials; and 30 per cent attended some meetings or training and dropped out before the projects ended. Lists of participant and non-participant farmers were drawn up with community key informants, and each list was randomly sampled, with the exception of the leaders, who were deliberately over-sampled. The term leader here refers to project participants selected for their early interest in sustainable agriculture, community leadership and communication qualities. In general, all of the leaders (ten) are included in the tables; but when necessary we checked whether removing them altered the statistical significance of comparisons. Table 5.2 shows the number of sampled leaders, normal participants, non-participants and control farmers.

Table 5.2 *Distribution of participants, non-participants and control farmers*

	Region 1 SJ de Flores	Region 2 Guaimaca	Region 3 Guacamayas	Total
Number of leaders	5	3	2	10
Number of normal participants	32	39	10	81
Number of non-participants	25	21	10	56
Number of control farmers[a]	16	15	0	31
Total	78	78	22	178

Note: [a]The original sample of the control communities was 35, but these included 4 project participants, who were treated as participant farmers, leaving 31 control farmers.

Data collection

After developing and testing the survey form, the household survey was carried out by a team of three to four enumerators (always including the second author) during May to June 2003. Since it was too early to ask about 2003, and 2002 was a drought year, most questions on agricultural practices related to 2001. Questions on maize related to the first rains crop, and those on beans to the second rains crop. Completed survey forms were checked on a daily basis.

Prior to the survey, a meeting was held in each community with a number of community representatives (usually 8 to 12). These meetings briefly covered community history; demography; projects and organizations; crop rotation and farm calendar; land tenure, labour, credit and market constraints; agricultural information sources; environmental issues; and the main household income sources.

To assess labour and economic constraints further, a crop budgeting exercise was carried out in November 2003 on a non-random sub-sample of 30 farmers, selected to represent key technology–crop combinations. This included 20 participants, 6 non-participants and 4 control farmers from four communities in regions 1–2.

Social and human capital

Finally, it is important to explain how we measured social and human capital. Our pragmatic interpretation of the term social capital can be thought of as the social interaction that accompanies technology development. Since we wanted to relate social capital to technology adoption, we decided to use a quantitative measure of social capital that could be assessed for each farmer. In the absence of a better measure, it was thought that farmer participation in projects and community organizations or institutions would be indicative both of a level of civic willingness and of the support (including agricultural information) that farmers were likely to receive from project and community-level interaction. We therefore derived four social capital indices for use in the analysis:[2]

1   *with-project* social capital: the full cumulative and current social capital index, including the numerical weight of participation in the LEIT projects;
2   *without-project* social capital: the current social capital index, but excluding the numerical influence of the LEIT projects;
3   *pre-project* social capital: the social capital index before the LEIT projects; and
4   *post-project* social capital: this excludes participation in the projects and attempts to show project impacts on social capital development.

We do not deny that these are crude and incomplete measures of social capital, as they take no account of the support from informal networks, kinship or community groups. They also make no attempt to include important contributors to community social capital such as trust and cooperation.

Human capital was analysed by reference to a range of individual variables, including age, education, the number of courses received, years of farming experience, family size and labour availability.

General results

General farming characteristics

Table 5.3 presents some farming-system characteristics in the project and control communities. The survey revealed an overall farm size of 5 hectares (ha), of which just over half was cultivated, with an average of 1.2ha of coffee, 0.85ha of basic grains and 0.14ha of vegetables. About half the farms had irrigation; most farmers with irrigation were those whose fields were sufficiently close to (and below) a spring or other water source and who invested in PVC piping and a simple sprinkling mechanism. About two-thirds grew coffee and a third grew vegetables. Other crops such as sugar cane, fruit orchards and trees for timber made up the balance. Two-thirds farmed their own land (87 per cent in the case of Guacamayas), as opposed to the use of rented, municipal or borrowed land; but less than 20 per cent had full title deeds. Half the farms used hired labour; two-fifths used herbicides on maize plots; and three-quarters used chemical fertilizers. A small proportion of farmers (15 per cent) used animal traction for preparing their maize plots. In the adoption analysis, we separated animal traction and manual cultivation farmers, as there were some substantial differences in farming practices.

Table 5.3 *General farming characteristics*

	Region 1 SJ de Flores (n = 78)	Region 2 Guaimaca (n = 78)	Region 3 Guacamayas (n = 22)	All (n = 78)
Farm size (ha)	4.3	6.4	2.4	5.0
Basic grains area (ha)	1.1	0.69	0.37	0.85
Vegetable area (ha)	0.07	0.18	0.27	0.14
Coffee area (ha)	0.49	1.7	0.53	1.2
Total chemical fertilizers (kg)[a]	141	177	40	145
Percentage working own land	62	68	87	67
Percentage farms with irrigation	42	55	59	51
Percentage using animal traction (maize)	22	10	0	15
Percentage income from off-farm sources	45	25	50	37

Note: [a]Between maize, beans and the main vegetable type grown.

Table 5.3 also shows considerable variations in farming-system characteristics between the regions, especially between the two lower altitude areas and Guacamayas. These reflect differences in population pressures, water availability, slope and access to market. Guacamayas (mean cropped area 1.7ha) was closest to Tegucigalpa, while Guaimaca (cropped area 3.7ha) was most remote. San Juan de Flores depended more upon basic grains production (but with lower maize yields), and had less irrigation, vegetables and coffee. Guaimaca had larger farms, more coffee and was more dependent upon farm income and employment than the other areas. Guacamayas had much smaller farms and plot sizes on its steep hillsides, depended less upon basic grains and, due to its market access and irrigation advantages, was more dependent upon vegetables (68 per cent of farms) and coffee. The relative proximity to Tegucigalpa also explains the higher proportion of off-farm income. Chemical fertilizer and herbicide use was much lower in Guacamayas, with several farmers producing and marketing organic vegetables and coffee, and no farmer sampled used animal traction.

Participant, non-participant and control farmers

A comparison was made of farmer characteristics in the project and control communities in regions 1 and 2 to check the representativeness of the project communities, while recognizing the limitations of the control community for region 1. The only significant differences were that control farmers had less *without-project* social capital, and a higher proportion grew coffee. There was little difference in human capital variables. Comparisons between participant and non-participant farmers in the project communities were more revealing. Table 5.4 shows the advantages of participants in terms of maize yields, irrigation, vegetable growing and social capital. When leaders were excluded, differences in maize yields and vegetable growing were no longer significant; but even so it is clear that participant farmers were better resourced than non-participants. An unanswered question at this point is how much these advantages were due to participation in the LEIT projects and utilization of their technologies.

Table 5.4 *Comparisons between participants and non-participants (regions 1–2)*

	Participants (n = 79)	Non-participants (n = 46)	Statistical difference
Vegetable area (ha)	0.16	0.08	<0.05
Maize yield (kilograms per hectare)	1105	806	<0.05
Total chemical fertilizers (kg)	163	145	No
Education (years)	2.7	2.3	No
Age	44.7	51.7	<0.05
Without-project social capital	2.6	1.2	<0.01
Percentage farms with irrigation	60	35	<0.05
Percentage farms with vegetables	43	22	<0.05

As previous comments imply, an important participant sub-group comprised the leaders. Since leaders were deliberately over-sampled, it is important to check for differences between them and normal participants (although statistical comparison is not possible). Table 5.5 reveals the advantages of leaders in terms of irrigation, vegetables, maize yields, and human and social capital. On the other hand, they had smaller farms, applied similar amounts of chemical fertilizers (but more organic fertilizers), and were similar in terms of coffee area and off-farm work.

Table 5.5 *Comparison between leaders and normal participants (regions 1–2)*

	Leaders (n = 8)	Normal participants (n = 71)	All participants (n = 79)
Vegetable area (ha)	0.22	0.15	0.16
Maize yield (kilograms per hectare)	1560	1053	1105
Education (years)	3.9	2.6	2.7
Without-project social capital	4.5	2.4	2.6
Percentage farms with irrigation	88	56	60
Percentage farms with vegetables	75	39	43

Enduring adoption levels of project technologies

One measure of enduring impact is the current adoption level of project-promoted technologies. Table 5.6 presents current (2001) adoption rates[3] of the main project technologies among all farmers surveyed, including control farmers. Since this was not a geographically representative sample, the results indicate only relative technology uptake. Regional comparison is also misleading in that there was no control community for Guacamayas. Figure 5.2 provides a fairer regional comparison in that it shows enduring adoption of the main LEITs in the project communities.

Table 5.6 *Adoption of project technologies (percentage of all farmers)*

	Region 1 SJ Flores (n = 77[a])	Region 2 Guaimaca (n = 78[a])	Region 3 Guacamayas (n = 17[a])	Total (n = 172[a])	Without leaders (n = 163[a])
Percentage in-row tillage (IRT) with maize	8	18	47	16	14
Percentage IRT with beans	6 (71)	21 (66)	58 (12)	17 (149)	14 (141)
Percentage IRT with vegetables	29 (17)	39 (33)	73 (15)	45 (65)	42 (57)
Percentage green manure (GRM) with maize	14	16	12	15	13
Percentage organic fertilizers on maize	9	8	47	12	10
Percentage organic fertilizers on vegetables	12 (17)	9 (33)	87 (15)	28 (65)	26 (57)
Percentage with live barriers	21 (78)	21 (78)	27 (22)	21 (178)	20 (168)

Note: [a]Unless otherwise specified.

Table 5.6 also reveals that IRT has been more enduring in conjunction with vegetables than with basic grains. In 2001, 16 per cent of basic grain farmers and 45 per cent of vegetable farmers used IRT on at least part of their plots, and 15 per cent of maize farmers used GRMs. Furthermore, most (70 per cent) IRT maize farmers also grew vegetables, and over one third of IRT adopters used the same plot for vegetables, maize and beans. Only one quarter of IRT farmers also used GRMs, which contrasts with the finding of a 1993 survey in the ACORDE area (Arellanes, 1994) that IRT was a 'first step' in LEIT adoption.

The use of IRT on both grain crops and vegetables is much higher in Guacamayas than in regions 1 or 2, as is the use of organic fertilizers. Several factors explain these differences. Most fields in Guacamayas are on quite steep slopes, and the use of the type of terracing offered by IRT is particularly appropriate. In addition, Guacamayas has better access to commercial vegetable markets (principally in Tegucigalpa) and a higher proportion of farmers grow vegetables. Land is in short supply and basic grains are often rotated on the same plots. Finally, earlier experience with marketing organic coffee has led some farmers into organic vegetable production, helping to explain the higher use of organic fertilizer (mainly purchased chicken manure) and the lower use of synthetic fertilizers.

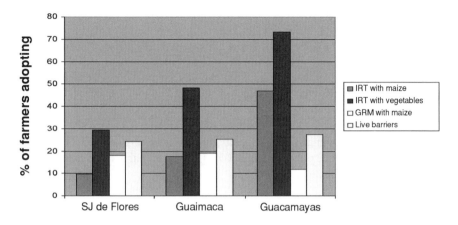

Figure 5.2 Adoption of technologies by region, excluding control farmers

Table 5.7 shows enduring adoption rates of the main LEITs by participant farmers (who were therefore exposed to similar amounts of information about the LEITs). For all four technologies in Table 5.7, over 80 per cent of enduring adopters in our sample were participants. Here we group regions 1 and 2 together. This again reveals the differences between Guacamayas (region 3) and the other areas. In the more typical project communities, about one in five participants were still using IRT with maize; almost half were cultivating vegetables with IRT; almost one quarter still cultivated GRMs with maize; and just under one third had live barriers. This is impressive when it is recalled that only about half of the 'participants' participated fully and to the end of the projects.

Table 5.7 *Adoption of low external-input technologies (LEITs) by participant farmers (including leaders)*

	Regions 1–2 (n = 79 [a])	Region 3 (n = 9 [a])	Total (n = 88 [a])
Percentage in-row tillage (IRT) with maize	20	78	25
Percentage IRT with vegetables	47 (34)	90 (10)	57 (44)
Percentage green manure (GRM) with maize	24	11	23
Percentage with live barriers	32	33 (12)	32 (91)

Note: [a]Unless otherwise specified.

Trends in adoption and abandonment

Current adoption rates disguise past interest in, and temporary adoption of, the focus technologies. About 90 per cent of participants and one quarter of non-participant and control farmers 'adopted' IRT and GRMs; the overall abandonment rates were 60 per cent for IRT and 76 per cent for GRMs. But these figures exaggerate adoption and abandonment since many farmers tested the technologies on a very small scale. Notwithstanding this, Figures 5.2 and 5.3 show adoption and abandonment of IRT (in any crop) and GRM (with maize) over time for each region.[4] This shows that post-project abandonment was steeper for GRMs than for IRT, and the gradual *post-project* IRT adoption in Guacamayas. The abandonment of IRT and GRMs in basic grain production is partially related to the problem of lower and more erratic rainfall in central Honduras over the past decade, and a general decline in basic grain production (R. Bunch, pers comm, 2003). GRMs, in particular, are vulnerable to drought conditions.

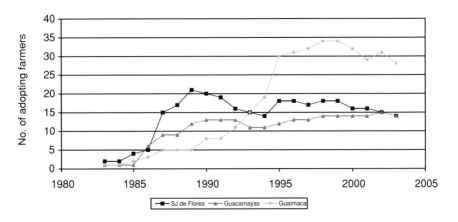

Figure 5.3 Tendencies in the adoption and abandonment of in-row tillage (IRT)
in the three regions

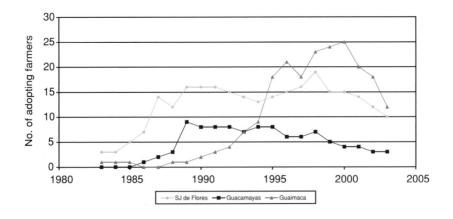

Figure 5.4 Tendencies in the adoption and abandonment of green manures
(GRMs) in the three regions

Why do some farmers adopt and others not?

Comparison of adopters and non-adopters

Various comparisons were made of plot and farmer characteristics of adopters and non-adopters in regions 1 and 2, including comparisons for each project technology; combining current IRT and GRM adopters and comparing them with non-adopters (see Table 5.8); and comparing IRT adopters (any crop) with those who abandoned IRT or never tried it (see Table 5.9). A logit analysis was also undertaken of IRT and GRM adoption, as well as for vegetable growing (see Annex 1).

Table 5.8 *Comparison of adopter and non-adopter farmers (regions 1–2)*

	Adopters (n = 44)	Manual non-adopters (n = 88)	Statistical difference	Animal traction non-adopters (n = 24)
Percentage farmers with irrigation	71	33	<0.01	59
Current age	43.2	48.5	<0.05	44.3
Years of education	2.5	2.3	No	3.3
Pre-project social capital	0.43	0.24	No	0.17
Without-project social capital	2.8	1.6	<0.01	2.1
Percentage farmers with vegetables	59	17	<0.01	27
Area of coffee and vegetables (ha)	2.0	1.2	No	0.98
Total chemical fertilizers (three crops) (kg)	195	95	<0.01	295

With regard to plot characteristics, both the simple statistical comparisons and the econometric analysis (for regions 1–2) found that irrigation was strongly related to adoption. In the econometric analysis, irrigation was a highly significant predictor of IRT adoption and a marginally significant (10 per cent) predictor of GRM adoption. By contrast, landownership,[5] plot size, farm size, slope and soil quality were not significantly related to adoption. With regard to the human capital variables, the most important factor in Tables 5.8 and 5.9 was age; adopters were significantly younger in the simple statistical comparisons but not in the econometric analysis. Associated with age, adopters were slightly less experienced, but had received more courses (partly due to the projects). No significant differences in education were found. Having removed the effect of the projects (without-project social capital), it was found that current social capital was significantly higher for adopters in the simple statistical comparisons. Pre-project social capital was also higher for adopters, but was not a significant factor in either the simple or econometric analysis.

Adopters were much more likely to grow vegetables and, linked to this, used significantly more chemical fertilizers than manual cultivation non-adopters. They had larger cash crop (coffee/vegetable) areas, but not significantly so. They were no less likely to work off-farm, implying that livelihood diversity was not a significant factor in LEIT adoption. While the few observations from Guacamayas preclude statistical analysis, adopters there were younger, better educated and had higher levels of social capital and larger cash crop areas.

Table 5.9 *Characteristics of farmers who never tried, abandoned and continue using in-row tillage (IRT) (project communities and participants in regions 1–2)*

	(a) Never tried IRT (n = 41)	(b) Abandoned IRT (n = 54)	Statistical difference (a–b)	(c) Continue using IRT (n = 32)	Statistical difference (c–rest)
Percentage with irrigation	22	58	<0.01	75	<0.01
Current age	52.5	46.4	<0.01	41.0	<0.01
Years of education	2.1	2.8	No	2.9	No
Pre-project social capital	0.20	0.41	<0.05	0.36	No
With-project social capital	1.5	2.9	<0.01	3.6	<0.01
Percentage farmers with vegetables	12	33	<0.05	69	<0.01
Area of coffee and vegetables (ha)	0.84	1.3	No	1.6	No
Total chemical fertilizers (three crops) (kg)	68	195	<0.01	209	No

As shown in Table 5.8, animal traction farmers were separated from manual non-adopting maize farmers. This was because of some important differences in socio-economic and farming-system characteristics from the manual non-adopters, such as better access to water, more years of education and a heavier use of chemical inputs by animal traction farmers. For example, animal traction farmers used twice as much chemical fertilizer per hectare as manual cultivation farmers (weighted average of adopters and non-adopters).

How important was resource wealth for adoption?

The above analysis suggests that LEIT adopters and vegetable farmers were, on average, better resourced. Access to irrigation was the main determinant of vegetable farming; farmers without water were likely to be in a state of relative poverty. The correlation between basic grain yields and fertilizer use (see Annex 2) also suggests that capacity to buy fertilizers was a key factor in family food production. These observations suggest the following tentative resource wealth classification of farmers in regions 1 and 2:

- 'least-resourced' farmers without irrigation and vegetables, and who applied no fertilizers to their maize (n = 32);
- 'poorly resourced' farmers without irrigation or vegetables, but who applied fertilizers on their maize plots (n = 40);

- 'better resourced' farmers with irrigation or vegetables, but not both (n = 42);
- 'best-resourced' farmers with irrigation and vegetables (n = 42).

Further analysis proved these categories to be highly consistent in terms of various possible wealth-related variables, such as farm size, cash crop area, basic grain yields, total fertilizer use, landownership, hired labour use, and dependence upon off-farm income.[6] Table 5.10 shows that there were no IRT adopters and three GRM adopters among the 32 'least-resourced' farmers, compared with 16 IRT adopters and 12 GRM adopters among the 42 'best-resourced' farmers.

Table 5.10 *Adoption according to resource wealth category (regions 1–2)*

Resource wealth category:	Least resourced	Poorly resourced	Better resourced	Best resourced
In-row tillage (IRT) maize	0%	13%	7%	29%
(n = 154)	(0/32)	(5/40)	(3/41)	(12/41)
IRT beans	0%	9%	6%	31%
(n = 136)	(0/28)	(3/34)	(2/35)	(12/39)
IRT vegetables	0%	0%	25%	38%
(n = 50)	(0/0)	(0/0)	(2/8)	(16/42)
Green manure	9%	10%	10%	29%
(GRM) with maize (n = 154)	(3/32)	(4/40)	(4/41)	(12/41)

This raises the question of the extent to which the current resource wealth status of adopters has resulted from the adoption of LEITs, or whether it predated and helped to explain adoption. This is an important discussion because if LEIT adoption were related to pre-existing income or wealth, it implies a constraint to adoption for resource-poor farmers. We found very few adopters among currently resource-poor farmers; but the 1993 study in Cantarranas found no correlation between farm income and adoption levels (Arrellanes, 1994).

The hypothesis that LEIT adoption is a determinant of wealth or income appears to be given some credibility by the econometric analysis of Jansen et al (2003), which showed hillside farm income to be correlated with soil fertility (and education). On the other hand, regressions on basic grain yields found that the LEITs were not significant, whereas fertilizer levels (possibly related to wealth) were highly significant (see Annex 2).

Possible clues to how much the projects contributed to the current welfare of IRT adopters are:

- whether IRT preceded vegetables; and
- whether more project participants with irrigation (compared with non-participants with irrigation) now grow vegetables.

We found that about 80 per cent of IRT vegetable farmers started growing vegetables during or since the projects, and 70 per cent of project participants with irrigation grew vegetables, compared with 44 per cent of non-participants with irrigation. Farmers in Guacamayas also felt that IRT was key to vegetable farming on their steep slopes. This suggests that IRT was an important stimulus to vegetable farming.

Arrellanes (1994) also tried to connect IRT adoption with vegetable growing, but found no clear connection between the two decisions, and concluded that farm income was probably the main determinant of vegetable production due to its high start-up costs (especially if this included installing irrigation). We checked for a relationship between coffee area, the main cash crop at the time of the projects, and vegetable growing. The econometric analysis (see Annex 1) revealed that (post-project) coffee area, education and irrigation were all highly correlated with vegetable growing. A further key factor was market access: Pender et al (2001) concluded that roads have provided the main stimulus to vegetable growing. In sum, we support Arrellanes's conclusion that income, market access and education were the main determinants of vegetable production, but suggest that IRT helped to facilitate it.

The balance of this evidence makes it hard to reach a clear conclusion on the role of 'resource wealth' in LEIT adoption and, thus, whether 'poverty' was a constraint to adoption. However, we were still concerned about the relatively low adoption rate by apparently poorer basic grain farmers. Field observations and discussions led us to consider the possibility that the labour cost of establishing IRT could be a problem for resource-poor basic grain farmers. While the study revealed that both IRT and GRMs were almost certainly labour-saving on an annual or maintenance basis, most farmers mentioned the importance of the initial labour costs of IRT.

In the crop budgets, 12 IRT farmers working in a range of conditions and slopes took an average of 67 days (a range of 41–103) per hectare to construct the mini-terraces.[7] About half of the labour was hired. The labour cost, including family labour, worked out at about US\$170 per hectare.[8] An important consideration here was whether the mini-terraces were constructed on a 'one-off' basis or gradually over the years. These farmers said that, although they developed IRT on their farms gradually in terms of area, development of the terraces was mainly in the first year. The difficulty of meeting this investment or of incurring the opportunity costs for farmers with limited cash income could partly explain the low adoption rate of IRT by resource-poor farmers and, thus, why poverty may be a constraint to IRT adoption.

Profitability, labour use and LEIT adoption

There is a consensus that if LEIT does not increase yields, its adoption will be limited (Bunch, 1999). The crop budgets presented in Annex 3 confirm:

- the higher maize yields and profitability of adopters;
- that vegetable production was much more profitable than basic grain production; and

- that with their lower costs and slightly higher yields (from a very small sample), IRT tomato producers obtained higher economic returns than non-adopters.

The higher economic returns to land, labour and capital under IRT cultivation, and, to a lesser extent, the use of GRMs, imply strong economic incentives for adoption. But this conclusion is premature for IRT since it takes no account of the costs of terrace establishment (not to mention the learning costs). A simple cost-benefit model was developed assuming an investment cost of about 70 days per hectare. Even assuming that the terraces would last only ten years, internal rates of return of about 90 per cent for IRT maize and 150 per cent for IRT tomatoes were estimated.

The higher adoption rate of IRT with vegetables, especially in Guacamayas, confirms the widespread view (e.g. Lutz et al, 1998) that profitability is a key determinant of LEIT adoption. The econometric study of Jansen et al (2003) also identified market access as a significant predictor of hillside technology adoption. Ruben (2001) argues that market engagement is a condition for LEIT uptake since the prospect of higher cash returns provides the willingness to invest in soil-improving technologies.

Another well-developed and empirically backed viewpoint is that a major reason for the low adoption of LEIT in Honduras and elsewhere is the problem of increased labour requirements and associated farmer opportunity costs (Pender et al, 2001; Hellin and Schrader, 2003). But in the survey, 80 per cent of IRT adopters and 60 per cent of GRM adopters opined that the technologies saved labour; only 11 per cent of IRT abandoners and 1 per cent of GRM abandoners cited labour as the cause of abandonment; and labour inputs recorded in the crop budgets were similar (maize) or less (tomatoes) for IRT farmers compared with non-adopters (see Annex 3). Therefore, our conclusion is that IRT and GRM were not more labour-demanding and were probably labour-saving.

The roles of social and human capital

Did pre-existing levels of social and human capital contribute to enduring adoption? While the mean *pre-project* social capital index was somewhat higher for adopters (see Table 5.11), the absolute levels were rather low, probably preventing significant differences (including in the econometric analysis).

Previous studies in Honduras identify community organization as an important determinant of technology adoption (Pender and Scherr, 2002; Jansen et al, 2003). One reason for this is that organizations can reduce individual learning costs. Both the survey data and community discussions revealed low levels of social capital before the projects, so the projects could not build on already strong local organizations.

The exception is Guacamayas, where the Comisajul Agricultural Cooperative was established a year before the ACORDE project through the initiative of a large landowner who lived near the community. The cooperative, which now markets organic coffee exports, encouraged technology adoption and group work. But community organization was only one of the many advantages of Guacamayas. The community has relatively good transport connections and proximity to the main national market. Its physical advantages include irrigation availability, rainfall patterns (compared with lower altitudes) and, for IRT adoption, the very steep slopes. It is adjacent to the La Tigra protected area, which has helped to develop environmental awareness. It is also distinguished by the number and quality of its local leaders. The project played a key role in identifying and further developing these leaders (Suazo, 2004).

Table 5.11 *Summary of social capital indices (regions 1–2)*

	Pre-project social capital	With-project social capital	Without-project social capital	Post-project social capital
Leaders (n = 8)	0.5	5.5	4.5	4.1
Participants (n = 71)	0.4	3.1	2.4	2.0
Non-participants (n = 44)	0.2	1.2	1.2	1.1
Control farmers (n = 31)	0.1	1.7	1.7	1.6
Adopters (n = 44)	0.4	3.0	2.8	2.3
Non-adopters (n = 88)	0.2	1.7	1.6	1.4

With regard to human capital, education was not significant in LEIT adoption; but it was highly significant for vegetable farming (also found by Arrellanes, 1994). It was therefore a probable determinant of household wealth (Jansen et al, 2003). In general, adopters were younger (although age was not significant in the econometric analysis). Neither was family size significant; this is consistent with the finding that the main LEITs were labour-saving since a larger family size among LEIT farmers might have implied a labour constraint to adoption. While the measurable human capital variables suggest that human capital did not have much role in adoption, a qualitative observation noted by the whole research team was that there were identifiable personality differences between adopters and non-adopters. Adopters seemed livelier and had more positive attitudes, especially to agriculture; this may also reflect their apparently superior resource wealth status. We therefore think that there were human capital differences, but not of a type easily captured by measurable variables.

A plausible view is that human and social capital are likely to be more important in explaining adoption when technologies are:

- more demanding in terms of their knowledge content and learning costs; and
- less profitable in the short term.

With regard to the first point, Arrellanes (1994), at least, felt that IRT and GRM were relatively easy to understand. In terms of less-profitable technologies, these technologies quickly lead to higher yields and do not have the slow payback problems of more traditional soil conservation technologies (Hellin and Schrader, 2003). A rather tenuous extension of this logic is that human and social capital were marginally more important for LEIT adoption associated with basic grains due to the lower profitability. The more obvious observation (at least for IRT) is that a potentially profitable crop (vegetables) greatly facilitated adoption.

Innovative farmers

An important aspect of sustainability is the capacity of farmers to innovate and adapt to the constant biological, technological and financial changes in the rural economy. In the survey, farmers were asked to describe any 'experiment' that they had carried out, and which involved use of a control plot. These included modifications or innovations to the project technologies, and other experiments – for example, with alternative remedies for pests and diseases, inter-planting and vegetable production. We did not include the introduction of new crops or simple variants of already established management practices as innovations.

Based on these definitions, we identified 22 farmers (12.4 per cent of the sample) who had conducted experiments. Half of these carried out two or more experiments; 73 per cent of these were project participants and the rest learned to do experiments from other projects. Of the participants, 20 per cent conducted experiments. About one third of these were leaders, so that 12 per cent of ordinary participants conducted experiments. In terms of innovation with, or modification of, the main project technologies, 19 per cent of adopters made significant modifications, mostly of IRT. There is, therefore, considerable evidence of experimentation and innovation, although without comparative data it is difficult and unhelpful to say whether these are relatively low or high levels from the perspective of enduring impacts. Limited evidence of 'proactive experimentation' by Honduran farmers has been noted by Humphries et al (2000).

Dissemination pathways and information flows

Almost all of the adopters of IRT (97 per cent) and GRMs (91 per cent), and all live barrier adopters in the project communities of regions 1 and 2 learned about the technologies as project participants. Only one IRT adopter and three GRM adopters said that they found out about the technologies from neighbours or that someone else had told them. In Guacamayas, 75 per cent of IRT adopters were participants.

There was more evidence of spontaneous adoption in the two control communities, even though this adoption was largely temporary; 57 per cent (20) of control farmers tried out IRT. Of these, 45 per cent learned about it through an extensionist or some other contact with the projects; 15 per cent learned about it from other projects; and the remaining 40 per cent learned from neighbours or because someone else had told them. In the case of GRM, 54 per cent (19) of control farmers tried it out: 53 per cent of these learned about it from projects (including in other agricultural projects) and 47 per cent observed it from neighbours or because 'someone' told them.

With regard to information diffusion, 90 per cent of IRT adopters and 64 per cent of GRM adopters in regions 1 and 2 said that they had shared information about the LEITs with others, mainly neighbours and relations. Overall, only about 20 per cent of survey respondents said that they did not know about the main project technologies. The dominant view from farmer and key informant discussions was that the main flow of agricultural information was extensionist to farmer and farmer to farmer, and that little exchange of agricultural information took place through community organizations unless these had a specific agricultural focus.

The limited spontaneous adoption of the LEITs, except in Guacamayas, should not, however, be ascribed to a failure of farmer-to-farmer extension methods. It is probably more related to the attractiveness of the technologies in situations of limited water availability for irrigation and to increasing incidence of drought. As already mentioned, IRT and GRM perform better with higher moisture levels. In contrast to some of the other project areas, a healthy flow of agricultural information between farmers was noted in the relatively water-rich Guacamayas (Suazo, 2004).

Project impacts on social and human capital

Did the project experiences stimulate subsequent social and human capital development? We found that participant and adopter farmers recorded significantly higher *without-project* and *post-project* social capital scores than non-participants and non-adopters, respectively (see Table 5.11). Thus, the projects have probably resulted in increased participation in community life, organizations, projects, church groups, etc. We use the word probably since other factors (including other projects) may have influenced these farmer scores, and also because the *pre-project* scores imply that some farmers already had a predisposition to community or project involvement.

There is also some qualitative, if sketchy, support for this finding. For example, some participants felt that the project experiences contributed to an increased interest in group work and encouraged them to participate in community organizations, and that before the projects it was uncommon for farmers to meet specifically to discuss farming methods. Some informal 'neighbourhood' farming groups, involving mutual support and some collective work, were formed with the encouragement and support of the leaders, but did not continue beyond the projects. On the other hand, the projects did not appear to result in increased farmer organization or networking, whether formal or informal.

We also tried to link the apparent levels of community trust and cooperation (based on comments by extensionists and community members) to project outcomes; but the few observations per community made it difficult to establish a clear pattern. The general viewpoint at the community meetings was that the higher levels of trust and cooperation in some communities were mainly due to the strength of local churches. The effect of reduced alcohol consumption, for religious reasons, on community harmony was regarded as an important factor. But on an individual farmer level there was no statistical relationship between religious activity (derived from a 'church activeness index') and technology adoption.

In terms of human capital impacts, again we think that the projects had positive impacts, but we have no hard data to prove this. Participatory observation and field discussions suggested that self-esteem had increased as a result of acquiring technical knowledge, increasing yields, sharing information, etc. While the projects may have helped to develop positive attitudes, it may also be that those who joined the projects and/or adopted the technologies were more open to new ideas in the first place.

Were the 'low external-input technology' (LEIT) adopters 'low-input' farmers?

A further interesting question is whether the LEIT adopters used less external or chemical inputs than other farmers. Table 5.12 shows that IRT maize farmers used more chemical fertilizers[9] (although not significantly more) and slightly fewer herbicides than traditional manual cultivation farmers. Animal traction maize farmers were heavier chemical-input users; there was also no evidence of substitution of chemical fertilizers by GRM adopters, and little difference between basic grain adopters and non-adopters in the use of organic fertilizers, pesticides and fungicides.

IRT vegetable farmers did use (non-significantly) fewer chemical fertilizers than non-adopters; IRT tomato growers used about one third less on average, and IRT cabbage growers about 40 per cent less. The crop budgets (see Annex 3) also revealed lower chemical-input levels by IRT tomato farmers. But there was little difference in the use of organic fertilizers or other chemical inputs. By contrast, in Guacamayas, most IRT vegetable farmers (carrots were the main crop) used organic fertilizers with very few chemical fertilizers, and some of them were growing certified organic vegetables.

The final column of Table 5.12 shows that adopters were more likely to try out alternative remedies for pests and diseases, although further analysis revealed that these farmers applied above-average levels of pesticides and fungicides. For observers such as Ruben (2001), it is unsurprising to find organic farming methods and inputs used as complements rather than substitutes for chemical inputs, since organic inputs can improve the efficiency of chemical inputs. For example, chemical fertilizer uptake is more efficient with better moisture conservation. Multiple regressions on basic grain yields (see Annex 2) revealed that fertilizer application rate was the main, and highly significant, determinant of yield. It appears that the adopters could better afford the yield-enhancing chemical fertilizers.

Table 5.12 *Use of chemical and organic inputs with/without LEITs (regions 1–2)*

	Chemical fertilizers (kilograms per hectare)	Percentage using organic fertilizers	Percentage using herbicides	Percentage using pesticides	Percentage using alternative remedies[a]
With LEITs:					
Maize with in-row tillage (IRT) (n = 20)	111	15	25	9	30
Maize with green manures (GRMs) (n = 23)	117	13	39	13	35
Vegetables with IRT (n = 18)	579	11	22	89	39
Without LEITs:					
Traditional manual maize (n = 109)	65	6	39	11	6
Animal traction maize (n = 23)	150	17	68	24	24
Maize without GRM (n = 131)	91	8	42	13	8
Vegetables without IRT (n = 32)	806	9	18	94	10

Note: [a]Alternative (to purchased pesticides and fungicides) remedies for pests and diseases.

Summary

This study revealed some important enduring impacts of the projects in terms of technology utilization and some evidence of increased social and human capital. For example, about half of the vegetable growers sampled in the more typical project communities continued to use IRT; about one in five participant farmers still grew maize using IRT; one in four still used GRMs with maize; and just over 30 per cent still had live barriers. In Guacamayas, 90 per cent of vegetable-growing participants and over 75 per cent of maize-growing participants used IRT. The high adoption rates of IRT and organic fertilizers in Guacamayas are related to a combination of factors, including access to water and markets; strong community organization; dynamic community leaders; steep slopes (favouring IRT); the development of environmental awareness; and the role of coffee in financing vegetable production.

As in other studies of the dissemination of soil conservation technologies in Honduras (e.g. Hellin and Schrader, 2003), adoption was mainly concentrated among project participants, and spontaneous adoption, often associated with the farmer-to-farmer dissemination model, was limited. Most enduring non-participant and control farmer adopters had some direct contact with the projects – for example, with project extensionists. This implies that more formal extension methods are still needed to complement farmer-to-farmer dissemination. The study also found that 12 per cent of ordinary participants and 20 per cent of all participants (including leaders) conducted experiments using control plots, and that 19 per cent of adopters innovated with the technologies by carrying out further adaptation.

In general, social and human capital did not have a significant influence on LEIT adoption, except in Guacamayas, where a strong community organization was formed just before the project. As in other Honduran adoption studies, education was not a significant determinant of LEIT adoption (but was highly significant for vegetable growing). Adopters were generally younger, although age was not significant in the econometric study. The limited role of social and human capital in LEIT adoption may be related to the relative simplicity of these technologies,and their rapid payback compared with more traditional cross-slope soil and water conservation methods.

The projects had positive impacts on current levels of social and human capital. Both project participants and adopters had significantly higher levels of *post-project* social capital. Adopters also seemed to have more positive attitudes to farming, possibly as a result of being involved in a successful project. The projects did not result in (or attempt) the development of formal farmer organizations. While there was no evidence of (project-induced) informal agricultural information networks, the specific technology orientation of the study meant that we did not pick up the flow of more general farming or market information that might be transmitted from farmer to farmer.

The study revealed that profitability and water availability were the most important LEIT adoption factors. For example, most (70 per cent) of IRT maize farmers (in all the communities) also grew vegetables. It appears that a profitable cash crop was very important for maintaining farmer interest. Perhaps surprisingly, land tenure and slope were not significant determinants of IRT uptake. Except for the labour establishment costs of IRT, we did not find that labour was a limiting factor for LEIT adoption; rather, the evidence was that these technologies were labour-saving in terms of their annual operations.

We found that project participation and adoption were skewed to *currently* better resourced farmers; but it was unclear whether these resource differences existed prior to the projects or were the result of project impacts such as LEIT adoption. There was evidence for both viewpoints, making it difficult to reach a conclusion on whether wealth was a significant factor in adoption. But the study did reveal a possible poverty or credit constraint to IRT adoption; the initial labour investment is difficult for farmers without a good source of cash income.

Finally, the basic grain adopters of IRT and GRM seemed to view the technologies as complementary to the use of chemical fertilizers, and not as a means of substituting them and reducing their capital costs. There was, however, some evidence of reduction of chemical fertilizers by IRT vegetable farmers, although chemical input levels for adopters were not significantly lower, except in the case of Guacamayas, where farmers have access to profitable markets for organic produce.

Annex 1: Logit econometric analysis

Statistical specifications

This annex summarizes the results of multivariate logistic regressions of the relationships between adoption of in-row tillage (IRT), green manure (GRM) and vegetable growing, and several independent variables or 'regressors'. The probability of adopting a given technology for each of the $i = 1,...,N$ farmers in our sample is modelled using a logit function:

$$Pr(A = 1|X) = \frac{\exp(\beta_0 + \beta_1 X_1 + \cdots + \beta_K X_K)}{1 + \exp(\beta_0 + \beta_1 X_1 + \cdots + \beta_K X_K)}$$

where $X = (X_1,... X_K)$ is a set of K independent plot and farmer characteristics hypothesized to affect the probability of adoption (A = 1). Since the β are not directly interpretable, we report two summary measures of the correlation between each X variable and adoption behaviour. The first is the 'odds ratio' (OR). The odds of adoption for a given set of plot and farmer characteristics can be defined as:

$$ODDS(X_i) = \frac{Pr(A = 1|X_i)}{Pr(A = 1|X_i)} = \frac{Pr(A = 1|X_i)}{Pr(A = 1|X_i)} = \exp(\beta_0 + \beta_1 X_{1i} + \cdots + \beta_K X_{Ki})$$

That is, the odds of adoption, for a given set of plot/farmer characteristics, equals the probability of adoption divided by the probability of non-adoption. The second summary measure shows the 'marginal effect' (ME) of a one-unit change of each independent variable (X_k) on the probability of adoption:

$$ME(X_k) = \frac{\partial Pr(A = 1|X = med(X))}{\partial X_k} \qquad \beta_K \times \frac{\exp(med(X)\cdot\beta)}{[1 + \exp(med(X)\cdot\beta)]^2}$$

where $med(X)$ denotes that the marginal effect is evaluated at the median value of each independent variable. For binary regressors (yes/no type variables), the marginal effect is defined as the incremental change in the probability of adoption between those with $X_k = 1$ and $X_k = 0$ (i.e. $Pr(A = 1|X_k = 1) - Pr(A = 1|X_k = 0)$).

Both OR and ME are non-linear functions of the underlying logit coefficients, $\beta = (\beta_1,..., \beta_K)$. Standard errors are computed by applying the delta method to the asymptotic variance-covariance matrix associated with the maximum likelihood estimates for β. An implication in a small sample like this is that the significances of β, OR and ME may not agree. Standard errors are based on the 'sandwich estimator', clustered on communities, and hence are robust to the presence of heteroscedasticity and arbitrary patterns of serial correlation across farmers within the same village.

Results

Tables A5.1 and A5.2 report the OR and ME summary measures for each independent variable. The regressions exclude farmers from Guacamayas, and the IRT regression also excludes draft animal preparation of maize plots. Given the small number of observations and the likelihood of relationships between the independent variables, these results should be treated with great caution, especially with regard to making causal statements.

Table A5.1 *Odds ratios in the three regressions (standard errors)*

	(1) Adoption of in-row tillage (IRT)	(2) Adoption of green manures (GRMs)	(3) Adoption of vegetables
Project participant	6.362 (2.885)***	5.203 (3.683)**	2.287 (0.973)*
Pre-project social capital	0.890 (0.158)	1.442 (0.311)˙	1.154 (0.261)
Farmer age	0.959 (0.025)	0.989 (0.019)	0.990 (0.023)
Education grade attained	0.976 (0.126)	0.722 (0.134)˙	1.343 (0.093)***
Tenure	1.415 (0.740)	1.350 (0.735)	1.338 (0.582)
Irrigation	2.989 (1.251)***	1.785 (0.521)**	6.978 (3.065)***
Coffee and vegetable area	1.061 (0.125)	0.985 (0.119	–
Coffee area	–	–	1.251 (0.109)***
n (population sample size)	133	156	156
Quasi-Log Likelihood	–57.78	–55.27	–68.29
Percentage classified correctly	77.7%	85.9%	78.9%

Notes: ***, ** and * denote significance at the 1 per cent, 5 per cent and 10 per cent confidence levels.

Table A5.2 *Marginal effects at sample median values of regressors*

	(1) Adoption of in-row tillage (IRT)	(2) Adoption of green manures (GRMs)	(3) Adoption of vegetables
Project participant	0.199 (0.057)***	0.151 (0.050)****	0.067 (0.034)*
Pre-project social capital	−0.022 (0.036)	0.058 (0.035)	0.016 (0.026)
Farmer age	−0.008 (0.004)*	−0.002 (0.003)	−0.002 (0.002)
Education grade attained	−0.004 (0.024)	−0.051 (0.036)	0.032 (0.011)***
Tenure	0.0591 (0.089)	0.043 (0.068)	0.029 (0.037)
Irrigation	0.249 (0.087)***	0.107 (0.057)*	0.375 (0.081)***
Coffee and vegetable area	0.011 (0.023)	−0.002 (0.019)	–
Coffee area	–	–	0.025 (0.009)***
Pr(A = 1|X = median)	0.249	0.196	0.126

Notes: ***, ** and * denote significance at the 1 per cent, 5 per cent and 10 per cent confidence levels.

Annex 2: Multiple regression analysis of basic grain yields

The household survey recorded consistently higher mean yields for basic grain adopters (see Table A5.3), although only in the case of IRT with maize were they significantly higher than for non-adopters (the small number of adopters generally prevented significant differences). Arrellanes (1994) also found considerably higher maize yields among IRT farmers. Vegetable yields were not recorded; given the range of vegetables being grown, these would have been difficult to compare.

Table A5.3 *Average yields of basic grains (regions 1–2)*

	Maize (kilograms per hectare)	Beans (kilograms per hectare)
With in-row tillage (IRT)	1274	988
Traditional manual cultivation	858	592
Animal traction	1047	715
With green manure (GRM)	1255	–
Without GRM	884	–

Standard linear multiple regressions on maize and bean yields were carried out to assess the impact of the technologies and other potential explanatory factors. The following independent variables were used: quantity of fertilizers applied per unit of land area; area planted; land preparation technology; use of green manures (maize only); use of organic fertilizers; land tenure; the *with-project* social capital index; irrigation; soil quality; and slope.

The overall coefficient of determination (r^2) was 0.37 for maize (significant at <0.01) and 0.26 for beans (not significant). In both cases, the fertilizer application rate was a highly significant determinant of yields (<0.01). The maize yield was also inversely related to plot size (<0.01) and positively related to the *with-project* social capital index (<0.05). The project technologies were not significant in explaining yield differences.

In sum, and as also found by Arrellanes and Lee (2001), small maize plots with higher fertilizer levels were significantly higher yielding than larger plots with lower fertilizer rates. Other observers have also found that the adoption of some LEITs is inversely related to plot size. Holt-Jiménez (2001), for example, argues that larger farmers can better afford to degrade the soil. But the current study and that of Jansen et al (2003) indicate that slightly better resourced poor hillside farmers choose to manage their plots more intensively than the poorest farmers, although they have access to similar amounts of land (usufruct land is not generally scarce except in places such as Guacamayas). It appears that resource-poor farmers are less able to substitute capital for labour and are therefore forced to cultivate larger basic grain plots in order to meet their family food needs.

Annex 3: Results of the crop budgeting exercise

Table A5.4 presents the results of the purposive sample of LEIT and non-LEIT farmers from four communities in regions 1 to 2, selected to represent important technology–crop combinations. Notwithstanding the small non-random sample, the following observations can be made:

- There was no difference in labour inputs between LEIT and non-LEIT maize farmers, although the former used much more hired labour (and less family labour).
- IRT tomato growers used less labour than other tomato growers, mainly at the land preparation stage (but this calculation does not factor in the labour establishment costs of IRT).
- While LEIT maize farmers used much more chemical fertilizer than non-LEIT maize farmers, IRT tomato growers used less than other tomato growers.
- LEIT maize farmers recorded much higher maize yields, about double in the case of IRT, for the first rains crop of 2003, resulting in more favourable gross margins. Most significant is the gross margin per person day. When compared with an average daily hired labour wage of about US$2.65 (for a six- to seven-hour day), LEIT maize farmers clearly did much better. But the return per labour day for non-LEIT farmers implies that production was still rational (in terms of the return to family labour). When labour was costed in, however, there were negative returns to land and capital for non-LEIT maize farmers.
- IRT tomato farmers also used fewer pesticides so that their total chemical input costs were much lower. With a slightly higher yield and much lower costs, IRT tomato growers were much more profitable. Given the expenditure involved in tomato growing, the return per dollar invested is important. While this shows that IRT growers had double the return of normal growers, a return of only US$1.10 per dollar invested implies that tomato growing, even with IRT, is only a marginal economic activity, as well as being subject to high risks.

Table A5.4 *Partial budget analysis of main technology–crop associations*[a] *(regions 1–2)*

	Maize with in-row tillage (IRT) (n = 8)	Maize with green manure (GRM)[b] (n = 5)	Maize without IRT/GRM (n =8)	Tomatoes with IRT (n = 6)	Tomatoes without IRT (n = 6)
Labour in land preparation (days per hectare)	19	18	19	55	74
Number hiring labour in land preparation	4	2	1	3	5
Labour in weeding (days per hectare)	21	22	26	97	102
Number hiring labour for weeding	5	3	0	2	5
Number farmers using herbicides	3	3	3	1	0
Total labour days per hectare	77	74	77	469	555
Hired labour days per hectare (all operations)	23	24	0.4	176	286
Number farmers using chemical fertilizers	7	5	2	6	6
Chemical fertilizers (kilograms per hectare)	130	221	20	1027	1430
Cost of chemical fertilizers (US$ per hectare)	72	93	20	624	771
Cost of insecticides/fungicides (US$ per hectare)	2	2	5	633	830
Total cost of chemical inputs (US$ per hectare)	74	95	25	1257	1601
Production of maize (kilograms per hectare)	1690	1235	845	–	–
Production of tomatoes (metric tonnes per hectare)	–	–	–	62	52
Gross margin per hectare (US$ per hectare)[c]	157	46	−52	2883	1574
Gross margin per labour day (US$ per day)[d]	4.47	3.24	2.71	8.82	5.47
Gross margin per US$ invested[e]	1.5	0.4	−2.0	1.1	0.5

Notes: [a]Data are presented in terms of US$, kilograms and hectares derived from local measures using the following conversion rates: 17 Honduran lempira per US$1; 1 quintal (100lb) equals 45.4kg; 1 manzana equals 0.7ha.
[b]Three of the five GRM with maize farmers also used IRT.
[c]The gross margin per hectare equals the value of production, less the cash inputs and labour costs (including family labour) divided by the area. For the value of production, the average annual purchase price of basic grains in the communities was used.
[d]The gross margin per labour day equals the value of production, less cash input costs, but without deducting the cost of labour, and divided by the total number of labour days.
[e]The gross margin per dollar of capital invested equals the gross margin per hectare divided by the cash input and hired labour costs per hectare.

Notes

1   The authors wish to thank, first of all, the farmers of Cantarranas and Guaimaca Municipalities for patiently providing the information for this chapter. They are also grateful for the econometric analysis conducted by Bryan Graham (PhD candidate, Department of Economics, Harvard University) and for the support, guidance and comments provided by Milton Flores (CIDICCO), Gabino López (COSECHA) and Roland Bunch (COSECHA/World Neighbors), as well as for the comments of Jon Hellin (ITDG) on the paper presented at the London Workshop, 22–23 June 2004. They also wish to thank their enumerators José Luis Gallardo, Xochilt Preza, Eduardo García, Tirza Maldonado and Ana Cecilia Romero. Several other people provided important additional information. This chapter is dedicated to the memory of their friend and inspiration, the late Don Elías Sanchez, founder of ACORDE and of the 'human farm' philosophy. The study was funded by the UK Department for International Development (DFID).

2   The social capital index was calculated as follows: each project (including the LEIT projects) or organization in which the farmer participated actively (according to the respondent) scored 1; moderately active participation scored 0.66; and low or passive participation scored 0.33. Where the respondent held a position in the organizations, an additional 0.5 points were added.

3   The minimum areas for enduring adoption were defined as 700 square metres for IRT and GRM with maize; 350 square metres for IRT with vegetables and beans; and 20m strips for live barriers.

4   Current adoption is slightly exaggerated in these figures, possibly due to the way in which farmers were asked about the technologies (there was an inconsistency in responses to one question about whether the farmer was currently using a technology, and to another question which asked when he/she adopted or abandoned the technology); but the graph is still useful for showing general trends.

5   Less than one fifth of farmers had full land title, so the comparison was based on a weaker concept of landownership, including locally recognized usufruct land rights, the predominant tenure category. Landownership was a significant factor in IRT adoption for Arrellanes (1994).

6   'Least-resourced' farmers were more dependent upon off-farm income. This could suggest that higher off-farm income compensated for lower on-farm income. But this was probably not the case, since most off-farm work was poorly paid farm work, and these farmers did not apply (could not afford?) yield-enhancing fertilizers on their basic grains.

7   In practice, the areas were much smaller than 1ha, but are expressed here on a per hectare basis for convenience.

8   This can be compared to Bunch's (1999) estimate of US$130 per hectare. Part of the disparity may be due to a difference in the amount of work done by hired labour; owners work faster and put in longer days.

9   This is not to say that LEIT farmers were using more chemical fertilizers (than non-LEIT farmers) per unit of output since our data revealed higher basic grain yields for IRT and GRM farmers (see Table A5.3).

6

Conservation by Committee: The Catchment Approach to Soil and Water Conservation in Nyanza Province, Western Kenya[1]

Catherine Longley, Nelson Mango, Wilson Nindo and Caleb Mango

Introduction

Efforts to promote low external-input technology (LEIT) rely upon the mobilization of local knowledge and labour to provide appropriate solutions to individual farm conditions. LEIT both draws upon and strengthens the skills and knowledge of farmers, as well as local social organization. By focusing on the Catchment Approach (1989–1999) of Kenya's National Soil and Water Conservation Programme (NSWCP), this chapter examines the extent to which LEIT has led to the adoption and spread of soil and water conservation (SWC) technologies and promoted the strengthening of human and social capital (i.e. farmers' skills, knowledge and organizations). The research was undertaken to examine differential responses to, and subsequent development of, the LEIT introduced. The study sought evidence on the degree to which the initiatives have contributed to strengthening farmer assets, allowing further adaptation and change beyond the initial project. Its findings suggest that the link between the Catchment Approach and any subsequent strengthening of social capital is somewhat tenuous.

After a brief description of the history of soil and water conservation in Kenya and the way in which the NSWCP was implemented, the study areas in Nyanza Province, western Kenya, are introduced. The question of who participated in the implementation of the SWC project is explored, and the degree of uptake of the different technologies promoted is analysed. Levels of adoption are examined in relation to various farmer characteristics, and the extent to which farmers adapted the technologies themselves is described. The closing sections of the chapter draw conclusions about the strengthening of farmer knowledge and the development of groups and organizations.

The evolution of soil and water conservation in Kenya

The colonial era (1930–1962)

The severity of the erosion problem in Kenya was recognized as early as the 1920s, and by the 1930s the government had already started formulating ways of conserving the land (Thomas et al, 1989). The colonial authorities addressed the problem of soil erosion by implementing district-level bylaws specific to 'African-held land' which focused on coffee and cotton – cash crops that had been forced upon farmers. Farmers were not allowed to plough steep land, cultivate along stream channels or clear forests. Contour farming, tree planting, terrace strip cropping and the de-stocking of herds were encouraged, and certain areas were closed off to prevent grazing (Tiffen et al, 1994). The local administration and agricultural officials rigorously enforced these stipulations and stiff penalties were imposed on farmers who failed to comply.

The lost years (1963–1972)

During the struggle for independence in the 1950s, resistance to coerced soil conservation became part of political agitation; pro-independence politicians campaigned strongly against compelling people to construct conservation structures. During the first decades of independence, continued government involvement in soil conservation became both politically and socially untenable. More terraces disappeared than were being constructed – hence the term 'lost years' (Holmberg, 1985). Soil erosion consequently increased, causing such concern that a land-use commission was set up in 1970 on the instructions of President Kenyatta to address the increasing degradation of natural resources in the entire country.

Revival of soil conservation (1972–1988)

By the time of the United Nations Conference on the Human Environment in Stockholm in 1972, land degradation was considered to be a major environmental problem in Kenya. In 1974, the National Soil and Water Conservation Programme (NSWCP) was launched with the aim of increasing and sustaining agricultural production by the introduction of simple, cheap and effective conservation measures to be carried out by farmers. The initial focus was on individual farms and farmers could choose for themselves the soil and water conservation (SWC) measures that they felt appropriate. During the 1980s, this approach gave way to Training and Visit (T&V) systems and then later to the Catchment Approach.

The Catchment Approach (1988–1998)

The way in which the Catchment Approach was implemented is described in the following section. It was felt that a strategy that addressed itself to the treatment of whole catchments would achieve results of a higher order, both quantitatively and qualitatively (Pretty et al, 1995; Harding et al, 1996). The benefits of the Catchment Approach were viewed in terms of the high visibility of the conservation effort; the continuous treatment of farms; the safe conveyance of excess runoff in the high rainfall areas and water harvesting in arid and semi-arid lands; the implementation of soil conservation activities all the year round; the involvement of neighbouring farmers in the decision-making and implementation process; and the development of a cadre of agriculturists specialized in soil conservation.

Focal area approach (2000–present)

In 2000, the National Agriculture and Livestock Extension Programme (NALEP) was launched. This programme retains the basic elements of the Catchment Approach (though catchments are now called focal areas) and follows a more demand-driven holistic approach in which a range of technical experts are involved. There is also encouragement of common-interest groups so that there is more group training and mutual support.

Background to the National Soil and Water Conservation Programme (NSWCP)

Organization and implementation of the programme at national level

The National Soil and Water Conservation Programme was implemented through the Soil and Water Conservation Branch[2] of the Land Development Division in the Department of Agriculture of the Ministry of Agriculture, Livestock Development and Marketing (Republic of Kenya 1996). The Swedish International Development Agency (Sida) agreed to provide financial and technical support, while the Kenya government undertook to allocate local resources and an enabling environment for conservation. The programme began moderately, but gradually expanded to cover more districts, mainly in the high-potential areas.

At the provincial and district levels, the programme was implemented by the provincial director of agriculture and district agriculture officers, respectively, through provincial and district soil conservation officers. At the divisional level, operations were carried out by the divisional planning teams (DPT), as described in the following section. Chiefs, assistant chiefs, district officers and district commissioners were recognized as potential collaborators in implementing soil conservation activities and were targeted for training through courses and seminars. The training of education professionals acted as the catalyst for including environmental conservation in the curricula of schools, colleges and universities, and for introducing soil and water activities into many primary and secondary schools.

The programme received considerable political support when President Daniel arap Moi ordered the introduction of 'soil conservation days' to be organized through the Office of the President. This was a manifestation of government commitment to soil conservation at the highest level.[3] President Moi also introduced a presidential trophy for the best catchment in the country, and he himself visited the winning catchment for any particular year.

Implementing the Catchment Approach at the local level

At the local level, the project was implemented by the divisional planning team (DPT), comprising the divisional soil conservation officer and two technical assistants. Given that catchments in the physical sense often included more than one village, the first task of the DPT was to delineate a catchment for project purposes. Although loosely based on a physically defined catchment, the project-defined catchment included between 200 and 300 households from villages or clans that were capable of working together. The team of three extension agents typically worked in three to four catchments each year. Priority was given to catchments where local people or administrations had asked for support, where erosion was serious or where the ministry had not worked before. Over the 26 or so years that the NSWCP was operational, the coverage of the project was so extensive that we had some difficulty in locating control sites within our study area (especially the high-potential areas) where the project had not been implemented. Although it was originally intended that the DPT should work continuously in each of the targeted catchments over the course of the year, in practice what tended to happen was that the DPT worked in each catchment consecutively, spending only three or four months in each. At the end of the implementation period, the DPT moved to another site and was generally too busy to make any subsequent follow-up visits unless this was part of another project (Agrisystems (EA) Ltd, 1998). Considerable emphasis was placed on reporting procedures, producing an impressive number of reports, maps and correspondence, much of which still exists in the divisional offices.

In each catchment, the DPT undertook a reconnaissance survey to demarcate the catchment, identify key stakeholders and undertake a participatory rural appraisal (PRA) exercise. A detailed map of the catchment was prepared showing the individual farm boundaries (the catchments in our sample varied in size from about 95 farms to 300 farms in total). Together with members of the catchment committee (see below), the agricultural extension agents were expected to visit every farm in the catchment area, recommend appropriate SWC structures for each farm, and indicate where these should be constructed by using stakes to 'lay out' the farm. This exercise was carried out quite thoroughly, with the majority of farms in a locality visited and laid out. The DPT recommended structures for each particular farm according to slope and degree of erosion, the presence of existing structures, the soil type and availability of materials such as stones, and their location was recorded in sketch maps.

Farmers were encouraged to construct the recommended structures in the course of subsequent visits by the extension agents and/or committee members. In each of the four catchment sites sampled, agricultural staff made up to six visits to the sample farms, though in 31 per cent of cases no visit at all was made. In 45 per cent of cases, either one or two visits, and in 24 per cent of cases, three or more visits were made. Over half of the sample farmers present in the village at the time of the catchment project attended between one and seven meetings with the agriculture staff.

Many of the activities undertaken by the extension agents were carried out with the assistance of members of the catchment committee. This committee was made up of between 8 and 15 men and women, often with a local technical assistant as *ex-officio* member. In one of our sample catchments, the *ex-officio* member (an extension agent from the local area who was subsequently transferred to his home location) played a pivotal role in ensuring the success of the Catchment Approach. According to the literature, these committees were sometimes derived from existing traditional or formal institutions (Pretty et al, 1995); but none of the four committees in our sample were formed from existing organizations. Members of the catchment committee were said to be elected; but in many cases their (s)election appeared to be strongly influenced by the local chief and his supporters, particularly in places where there were two or more clans or communities within the catchment. In such cases, the chief ensured a balance of clan and community representation on the committee to avoid potential problems. All of the committee members tended to be members of existing organizations at local level and their credentials and leadership qualities were therefore well known.

Training (lasting up to a week) was provided to the committee members on soil and water conservation, and it was expected that the committee would then train other farmers. Training techniques included public meetings (*barazas*), individual farmer and group extension methods, agricultural shows, study tours and excursions to areas where conservation programmes had been successfully implemented, and residential courses at farmers' training centres. In the study sites, the role of the committee members was understood to include the following: linking the agricultural extension officers with the catchment farmers; acting as a role model to other farmers; identifying the worst eroded farms and footpaths for appropriate action; and assisting extension officers in laying out farms.[4] Place et al (2003) report that such committees tended to perform poorly in providing information to other farmers (as opposed to members of other committees). As will be shown, our data results reveal a similar finding.

In general, the tools required for laying out and implementing soil conservation structures (for example, hoe, spade, hammer and *jembe*) were issued to committee members, though in two of our four sample catchment sites, the promised tools were never delivered. In some cases, Napier grass and aloe plants were also provided and the committee maintained bulking plots. In order to hasten the success of the Catchment Approach, ministry staff reported that incentives were sometimes handed out in the form of cash and T-shirts. Cash payments were said to be provided to allow farmers to extend certain structures such as cut-off drains and waterways beyond their own farm boundaries; but none of our sample farmers or committee members received any such payments. Although such incentives could be considered as subsidies, they acted as 'the carrot' in place of 'the stick' used during the colonial period. In some cases, these incentives motivated farmers to implement conservation measures in a collective manner. In general, however, the provision of incentives led to problems and was later phased out.

The technologies promoted

A wide range of SWC structures was promoted by the Catchment Approach, including physical (e.g. terraces, ditches, stone lines, cut-off drains and retention ditches), biological (e.g. grass strips, aloe strips, hedges and banana lines), and cultural practices (e.g. trash lines, mulching and fallow). The six most popular structures are described here.

Vegetative strips

Grass strips were the most popular structure to be implemented by our sample farmers. A grass strip is a narrow band of grass planted on cropland along the contour. The grass strip may be planted to a fodder grass such as Napier or left unploughed with self-seeding grasses, although this may aggravate the problem of weeds on the farm. The strips are usually about 0.5m–1m wide and spaced according to the slope of the land; on a slope of 27 per cent the width between strips should be 15m (Thomas et al, 1997). Grass strips are effective soil conservation measures on soils with good infiltration and slopes up to about 30 per cent. They are less labour-intensive than terraces. They should be allowed to develop a thick basal growth to slow down erosion and retain soil material. In semi-arid areas, sisal and aloe strips are commonly used because they are more drought-resistant. Establishing grass, sisal and aloe strips is relatively cheap and does not take much space on the farm.

Unploughed strips

These are sections within the farm running along the contours that are deliberately left unploughed in order to reduce the speed of surface runoff and enhance water infiltration. They are normally 1m wide and form ridges after some time as a result of the accumulated silt. The distance between the strips depends upon the angle of the slope; like grass strips, on a slope of 27 per cent the width between strips should be 15m (Thomas et al, 1997). The formation of unploughed strips involves no labour. Rather, their continued presence depends upon clear instructions to/supervision of those leading the oxen teams that tend to be hired for ploughing. A number of farmers complained that these hired teams start work very early in the morning, without waiting for instructions or supervision from the farmer.

Fanya juu and fanya chini terraces

A *fanya juu* terrace is made by digging a trench and throwing the soil uphill to form an embankment (*fanya juu* literally translates as 'do it up'). The cross-sectional profile of a *fanya juu* comprises an embankment to impound water, soil and nutrients; a storage area above the embankment to prevent overspill by runoff; and a berm or ledge to prevent the embankment soils from sliding back into the trench. Perennial grasses are planted on the embankment to stabilize it. The trench below the embankment may or may not be retained.

These terraces are applicable in areas where soil is too shallow for level bench terracing and on moderately steep slopes (e.g. below 20 per cent). The recommended width of the terrace is 0.5m and the depth is also 0.5m; when the width of the embankment is added, the terrace takes up more space than a grass or unploughed strip. It is quite labour-intensive to construct and requires an estimated 8.3 labour days to excavate 100m (Thomas et al, 1997).

A *fanya chini* is like a *fanya juu*, except that the soil dug from the trench is put on the lower side of the trench (*fanya chini* literally translates as 'do it down'.) The resulting embankment can be used to grow fodder. *Fanya chini* are less labour-intensive than *fanya juu,* but they do not lead to the formation of a bench terrace over time.

Retention ditches

Retention ditches – also called infiltration ditches – are large ditches that are designed to catch and retain all incoming runoff until it infiltrates the ground. Retention ditches are commonly used as an alternative to diversion ditches if there is no place to discharge runoff or if there is a need, as in semi-arid areas, to harvest water – for example, for bananas. Retention ditches are constructed along the contour, with closed ends; they are wide enough (typically 0.5m wide at the bottom and 1.5m wide at the top) and deep enough (0.5m) to hold all the runoff expected. In dry areas they can be useful for water harvesting on low slopes. They are often found on steep slopes in humid areas under small-scale farming where there is no opportunity to discharge runoff to a waterway. Retention ditches can be useful where soils are permeable, deep and stable, but are not suitable on shallow soils or in areas prone to landslide. Construction of a retention ditch is labour-intensive; the channel is excavated and the soil thrown on the lower side to form an embankment. The embankment should be at least 0.2m wide from the edge of the channel, thus sacrificing considerable space on the farm.

Cut-off drains (diversion ditches)

Cut-off drains are graded channels with a supporting ridge or bank on the lower side. They are constructed across a slope and are designed to intercept surface runoff and convey it safely to an outlet such as a waterway. Cut-off drains are used to protect cultivated land, compounds and roads from uncontrolled runoff, and to divert water away from active gully heads. Because of the large quantity of water that such structures are intended to convey, they are usually trapezoidal in shape and have a large capacity. A typical cut-off drain should be around 0.5m deep and 1m–1.5m wide. It takes a lot of space in the farm and is quite labour-intensive to construct; in some cases, it can be excavated by a machine.

Background to the case study area and the fieldwork sites

The seven study sites included four former catchments, as well as three control sites (where the Catchment Approach had not been implemented), sampled from high- and low-potential areas (see Table 6.1). The control sites were located in areas neighbouring the catchment sites – sometimes in neighbouring villages, as in the case of Kaol and Wang Adonji – and thus have similar agro-ecologies. The distinction between high- and low-potential areas is clearly defined within the agro-ecological classification system used in Kenya. As illustrated in the following sections, high-potential areas have higher rainfall and better soils than low-potential areas and thus have greater potential for crop agriculture. Key features of each of the seven sites are shown in Table 6.2.

High-potential areas

The study sites within the high-potential zones were Nyahera and Ojwando catchments and Nyabola control site. Within the high-potential zone, altitude ranges from 1355m to 1650m above sea level, and annual rainfall is 1100mm–1600mm in a bimodal pattern (the long rains are from March to June and the short rains are from August to October). The area is generally hot and wet with temperatures ranging from 14° Celsius (C) to 28° C, with an annual mean of 21°C. The soils range from sandy loam/sandy clay to red volcanic soils.

Living in a high-potential area, people grow a wide range of crops, the main ones reported by our sample farmers being maize, beans, bananas, groundnuts, sweet potato and sorghum. Maize and groundnuts were the main crops sold by our sample farmers, although coffee, tea and sugarcane form the main cash crops in other parts of the high-potential zone.[5] The average size of total land holdings was 2.4ha, 91 per cent of farmers used fertilizer, and about one quarter used hybrid or certified seed. Maize provided the main food staple for almost all of our sample households in the high-potential zone, survey results giving an annual mean household production of 974kg.[6] Households estimate that their grain production lasts an average of 8.5 months, with almost half of the sample farmers (43 per cent) in the high-potential sites never having to buy grain. Although cattle ownership is not high,[7] people in high-potential areas are more likely to own high-grade exotic dairy cattle than those in low-potential areas.

Data on household incomes clearly reveal that crop sales are considerably more important to households in the high-potential areas than in low-potential areas. Table 6.2 shows that 83 per cent of households in the high-potential areas listed these among their top three main income sources, providing on average 38 per cent of total household income. The way in which farming is valued by the survey respondents was determined by asking what advice they would give to a young person – that is, whether to become a full-time farmer, a part-time farmer or not a farmer. About one third of respondents in the high-potential areas were of the opinion that full-time farming was the best advice (mainly because the area is productive and cash crops can be grown), and two-thirds thought that part-time farming was best (to supplement other forms of income).

Table 6.1 *Key features of the study sites*

	Name of study site	Key features
High-potential area	Ojwando	Main livelihood strategy is crop farming and trading in agricultural produce. Napier grass had been introduced prior to the catchment project by the government Livestock Development Programme. The International Centre for Insect Physiology and Ecology (ICIPE) and the Kenya Agricultural Research Institute (KARI) have carried out research into crop pests and *Striga* spp control since 2000. Two church-based agricultural projects have worked in the catchment area. The Catchment Approach was implemented in 1998–1999
	Nyahera	Similar to Ojwando. In addition to crop farming, livestock keeping and livestock trading, about one quarter of the population are engaged in gold mining. There is a strong colonial influence on soil and water conservation (SWC). Most of the farm terraces date back to the colonial period. CARE-Kenya promoted agroforestry within the catchment during the late 1980s, and the Livestock Development Programme promoted Napier grass in the early 1990s. KARI undertook varietal research in 2002. The Catchment Approach was implemented in 1996–1997
	Nyabola (control)	Similar agro-ecology to Ojwando. Both CARE and the Forestry Department promoted tree planting during the early 1990s. The Livestock Development Programme promoted Napier grass during the late 1990s
Low-potential area	Wang'Adonji	Located near the lake. Farming/livestock, mat-making, trade and fishing are the primary means of livelihood. Soils are stony and sandy, and maize frequently fails due to long dry spells. Many farmers own farms located outside the catchment area. Massive runoff occurs during the rains due to low infiltration. The local agricultural extension agent comes from the community and is resident within the catchment area. Both KARI and CARE worked in the catchment during the late 1990s. The Catchment Approach was implemented during 1997–1998
	Lower Nyabola	Cotton and sorghum are cultivated, but livestock are more valued. Fishing also provides important livelihood strategy. Soils are sandy clay, and deep, wide gullies are clearly visible. The catchment is located close to the agricultural extension office, and the catchment committee chairman is a retired extension officer. Many local groups existed before the project. CARE-Kenya worked with women's groups to promote rice production and business enterprise. The Catchment Approach was implemented during 1998–1999

Upper Konditi (control)	Similar agro-ecology to Lower Nyabola. Sand harvesting is a major activity along the two seasonal rivers (a sign of heavy runoff emanating upstream), and there is evidence of massive surface runoff with rills and small gullies forming across farms. During 1986–1987, the Winam Gulf Project was active in the area and paid people to construct SWC structures. CARE-Kenya worked in the area in 2002
K'Aol (control)	A good number of farmers have other major farms located outside the village. A few individuals engage in fishing-related trading. Kaol is the smallest study site with about 30 households. The only external agencies to work in the area have been the Ministry of Agriculture and KARI

Table 6.2 *Three most important sources of household income during the past year*

Sources of income	High-potential area (n = 54 households)	Low-potential area (n = 74 households)
Sale of crops	44 (83%)	39 (53%)
Sale of livestock/products	29 (55%)	38 (52%)
Business or petty trade	28 (53%)	53 (73%)
Formal employment	7 (13%)	12 (16%)
Self-employment	4 (8%)	6 (8%)
Employment as casual labour	22 (42%)	24 (33%)
Remittances	24 (45%)	47 (65%)

Low-potential areas

The low-potential study sites are Lower Nyabola and Wang'Adonji catchments, and Upper Konditi and K'Aol control sites. The altitude of the low-potential area is 1140–1450m above sea level, and rainfall averages 1000mm per year. The long rains fall between March and April, and the short rains (in the years when they fall) occur between September and November. Due to the unreliability of the short rains, farmers tend to plant only in the long rains, when the weather is normally warm and semi-humid, whereas the short rains are characterized by hot, dry temperatures. The annual mean temperature is 34°C. The soils vary from sandy, sandy clay and clay to shallow young soils of mainly murram or gravel. The lowland areas close to Lake Victoria are prone to seasonal flooding.

The main crops grown by our sample farmers are maize, sorghum, groundnut, cowpeas, beans and vegetables. Crops do not provide a major source of income in the low-potential areas, but cash crops include cotton, groundnuts and sisal. The average size of total land holdings was 1.3ha, none of the sampled farmers used fertilizer, and about one quarter used hybrid or certified seed. Maize provides the main food staple to 64 per cent of sampled households; 36 per cent regard sorghum as their main staple. Based on farmers' estimates, annual mean maize production was 202kg and annual mean sorghum production was 152kg. Households estimate that their combined grain production lasts an average of 4.7 months. The majority of sample farmers in the low-potential areas (77 per cent) reported having to buy grain every year. With the exception of chickens, farmers in low-potential areas tended to own more livestock (cattle,[8] goats and sheep) than those in high-potential areas.

Data on household incomes show that crop sales are considerably less important in the low-potential areas: 53 per cent of households listed them among their main income sources (see Table 6.2), and crop sales provide only 14 per cent of total household income. By contrast, business or petty trade and remittances are far more important in the low-potential areas: these are among the main sources of income for 73 per cent and 65 per cent of households, respectively, accounting for an average of 31 and 19 per cent of total household income. In terms of the way in which people value farming as a means of livelihood, only about 10 per cent of respondents in the low-potential areas considered that full-time farming was best (where a water source was available for irrigation), whereas almost 90 per cent felt that part-time farming was best (to supplement other income) because buying food is expensive and rainfall is unreliable.

The study sites and sample farmers

The study sites fall within Nyando and Rachuonyo districts of Nyanza Province, as shown in Figure 6.1. The Luo, from the Nilotic language group, inhabit the two districts, having migrated from the north and west between three and five centuries ago. The Luo tend to reside in the low-potential areas, whereas the Kisii are predominant in the high-potential areas. All of our study sites were deliberately selected from among the Luo-speaking areas; the small number of Luo settlements in the high-potential areas border on Kisii-speaking communities.

Given the fact that the Catchment Approach attempted to work with all farm households within a catchment area, purposive sampling was used to include three different types of farmers within the catchment sites: committee members, non-committee members with laid-out farms, and farmers with non-laid-out farms – that is, those who were present at the time of the project but did not have their farms laid out or those who started farming in the area *after* the project (e.g. new households that had emerged or farmers who had immigrated or returned to the area).[9] Data were collected from a total of 128 farmers, as shown in Table 6.3. Both committee members and those with non-laid-out farms are over-represented in our sample. Although this limits precise statistical statements, it allows us to understand better who did or did not participate in the project. 'Project farmers' are those in the former catchment sites who are either committee members or non-committee members with laid-out farms. 'Non-project farmers' are those with non-laid-out farms, plus those from the control sites where the Catchment Approach was not implemented.

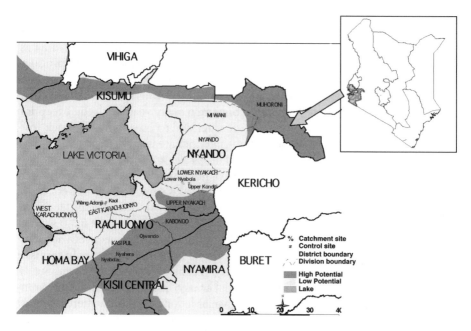

Figure 6.1 Location of study sites in Nyanza Province, western Kenya

Table 6.3 *Sample farmers*

| Name of study site | Project farmers | | Non-project farmers | | Total sample farmers |
	Committee members	Farmers with laid-out farms	Farmers with non-laid-out farms	Control farmers	
High-potential area					
Ojwando	6	9	3	0	18
Nyahera	4	8	6	0	18
Nyabola (control)	0	0	0	18	18
Low-potential area					
Wang'Adonji	6	6	6	0	18
Lower Nyabola	5	10	3	0	18
Upper Konditi (control)	0	0	0	20	20
K'Aol (control)	0	0	0	18	18
Total	21	33	18	56	128

Despite the high proportion of laid-out farms (90 per cent) in our sample catchments and the encouragement from the DPT and the catchment committee, farmers did not necessarily construct the recommended conservation structures immediately. Table 6.4 (based on information recorded by the DPT in the same year that the project was implemented) suggests that, across the four sample sites, just over half of the farms laid out were subsequently 'implemented'.[10] However, as will be seen below, many farmers constructed SWC structures on their farms a year or two after the project had been completed. In addition to the structures that were recommended on the individual farms, a number of structures were recommended across several farms and these were to be constructed communally. In the four catchments that we visited, approximately half of the farmers sampled also provided labour for the preparation of communal SWC structures.

Table 6.4 *Number of farms in sample catchments*

Catchment name	Approximate number of farms	Farms laid out	Farms actually implemented by the end of project
Ojwando	300	285	40
Nyahera	95	93	59
Wang'Adonji	300	260	186
Lower Nyabola	200	173	156

Source: Division Agriculture Office records

Who participated in the Catchment Approach?

There are not many differences between catchment committee members and other people in the catchment villages in terms of income sources, farm size, gender, household labour availability,[11] wealth[12] or education (see Table 6.5). Committee members certainly do not constitute an elite as their farms are somewhat smaller than those of their neighbours, most (75 per cent) have only primary education or no education at all, and none are classified as wealthy. What is significant, however, is that they tend to belong to a larger number of organizations than other farmers and they are more likely to have fields that are steeply sloped.

The committee members included in our survey were not rewarded in any major way, except for Wang'Adonji catchment where they were given T-shirts printed with the slogan 'Conserve Our Soil'. In two of the sample catchments, committee members remarked that they were mocked by other farmers who repeatedly questioned them as to why they were working for free. In Lower Nyabola, the catchment project was combined with a Food for Work programme, which was being coordinated by the Ministry of Agriculture and the provincial administration. The construction of SWC structures was the form of work that people had to undertake in order to receive the relief food. In Nyahera and Ojwando catchments, there were no incentives. The principal motivation for farmers to participate in the project was that their farms were either already experiencing high soil erosion or that the nature of their farms was prone to erosion (steep slopes).[13] Other reasons mentioned included previous knowledge of SWC measures (either as a contact farmer or during the colonial period) and a desire to benefit from SWC measures.

Table 6.5 *Characteristics of committee members compared with other catchment farmers*

Characteristic	Committee members (n = 21)	Other catchment farmers (n = 51)	Statistical significance
Percentage income from crops	34	30	ns
Percentage income from business or trade	19	24	ns
Total holding (acres)	3.0	4.3	ns
Percentage with steeply sloped field	60	26	<0.05
Percentage female	38	42	ns
Percentage with low household labour availability	27	15	ns
Percentage wealthy (wealth ranking)	0	18	ns
Number of memberships in organizations	2.1	1.3	<0.05
Percentage with some education	75	82	ns

Note: ns = not significant.

The levels of participation varied between individuals and between catchment sites. In Lower Nyabola catchment, for example, committee members worked with the extension agent, and a total of 173 plots were laid out for the soil and water conservation project (see Table 6.4). The committee held about eight meetings with the extension agent and organized several meetings with farmers; however, the farmers' attendance was poor. In general, throughout all the sample catchments, meetings with the catchment committee were not well attended: only 37 per cent of sample farmers present at the time of the project reported having attended one or more meetings with the committee, suggesting that the committees were either not very effective in mobilizing farmers or that the farmers were less motivated to attend.

Those from the catchment sites who did not participate in the project (those who did not have their farms laid out) are not statistically different from the project farmers, although there is a tendency for the non-project farmers to draw slightly more on non-farm income, to have less steeply sloping fields, to include fewer female farmers and to participate in fewer organizations.

Uptake of the different technologies

In general, the uptake of SWC practices would appear to be quite high in the sample sites; SWC structures were visible on 70 per cent of the 128 plots visited. Some of these structures, however, were implemented prior to the project. Over half (56 per cent) of sample farmers, in fact, implemented new SWC structures either during or after the project period.

A simple count of the types of structures recorded for the sample plots and when they were first established shows that many farmers in the study sites (both catchments and controls) were already practising SWC prior to the

Catchment Approach (see Tables 6.6a and 6.6b). The figures from previous periods can only be treated as rough approximations, but they give an idea of the extent and type of SWC activity before the project. The data from the project period are more accurate, and provide a picture of the extent and types of SWC technology used. As the introduction to this chapter makes clear, Kenya has had a long history of SWC activity, so it is perhaps not surprising to find that almost half (45 per cent) of the different structures recorded for the sample plots were reported to have been established prior to the Catchment Approach.

When we compare the types of structures for each period, there is clearly an increase in the diversity of types of structures over time. Those established during the colonial period were predominantly *fanya juu* and unploughed strips. With the advent of the NSWCP[14] (but prior to the Catchment Approach), unploughed strips remained popular, and farmers also began constructing grass strips, cut-off drains, hedges, tree lines and sisal strips. When the Catchment Approach was implemented in the case study catchments, the number of different types of structures established increased to over 20 to include those previously implemented, as well as retention ditches, grass/Napier plots and trenches.

Table 6.6a *Period of implementation of soil and water conservation (SWC) structures in high-potential sites*

Technology	Catchment farmers (n = 36)				Control farmers (n = 18)			
	Colonial period (1930–1962)	Prior to project (1963–1996/1997)	Post-project (1997/1998–2003)	Total	Colonial period (1930–1962)	Prior to project (1963–1996/1997)	Post-project (1997/1998–2003)	Total
Grass strips		5	24	29		1	6	7
Unploughed strips	1	4	4	9	3	3		6
Retention ditches and terraces [a]	6	2	7	15	1		3	4
Sisal/aloe strip			1	1				0
Vegetative lines (banana, hedges, trees)	1	5	3	9				0
Cultural practices [b]			2	2				0
Cut-off drains		3	1	4				0
Stone lines				0			2	2
Other		1	4	5		2	2	4
Total	8	20	46	74	4	6	13	23

Notes: [a] This includes fanya juu and fanya chini.
[b] This includes fallows and trash lines.

Table 6.6b *Period of implementation of SWC structures in low-potential sites*

Technology	Catchment farmers (n = 36)				Control farmers (n = 38)			
	Colonial period (1930–1962)	Prior to project (1963–1996/1997)	Post-project (1997/1998–2003)	Total	Colonial period (1930–1962)	Prior to project (1963–1996/1997)	Post-project (1997/1998–2003)	Total
Grass strips			3	3				0
Unploughed strips	3	6	5	14	1	3	5	9
Retention ditches and terraces [a]		2	3	5	2	2	3	7
Sisal/aloe strip	1	6	4	11			1	1
Vegetative lines (banana, hedges, trees)	1	5	1	7	1	2	4	7
Cultural practices [b]		1	1	2				0
Cut-off drains		1		1		2	4	6
Stone lines	1	1	1	3				0
Other			3	3				0
Total	6	22	21	49	4	9	17	30

Notes: [a] This includes fanya juu and fanya chini.
[b] This includes fallows and trash lines.

A total of 23 types of SWC activity were uncovered during the survey, and these have been classified into eight groups. Tables 6.6a and 6.6b indicate that approximately two-thirds of the farmers in the high-potential catchment sites established at least one grass strip on their farms in the post-project period and that about one third of farmers in control sites also established grass strips. The next most popular technology in the high-potential sites comprised various types of ditches or terraces; but – in contrast to results reported by Tiffen et al (1994) for Machakos – only a minority in the catchment villages established one of these. In the low-potential sites, uptake of SWC technology was much lower. The most widely adopted structures in the low-potential sites were unploughed strips and sisal or aloe strips, followed by grass strips and ditches. In the low-potential catchment sites, less than half of the farmers established one or more of such technologies (and less than one quarter of the farmers in the control sites). These differences are explored further below.

There are a number of reasons why grass strips are preferred by farmers. They do not require much labour to construct and compared with other technologies such as terraces or ditches, they do not take up too much space. In addition, the Catchment Approach was introduced more or less at the same time as zero-grazing dairy farming was being promoted in Kenya,[15] when Napier grass was being promoted as a fodder crop for dairy cows. Even farmers who did not own dairy cows were encouraged to grow Napier grass as a cash crop for sale to other farmers with exotic breeds of dairy cattle.

Figure 6.2 illustrates the temporal implementation of SWC structures for the period a few years prior to and after the catchment project in the different sample sites.[16] Although farmers' memories of the specific year in which they established their structures may not be entirely accurate, the general pattern clearly shows an increase in the number of structures implemented in the catchment sites at the time of the project, and this is mirrored by a small increase in the implementation of structures in the control sites about three years later. Although it is difficult to confirm, this could be explained by the spread of technologies from the catchment sites to the neighbouring control sites. It could also be explained by the influence of other projects. In both catchment and control sites, the implementation of structures continued at a slightly higher rate in the years after the project than during the years immediately before the project.

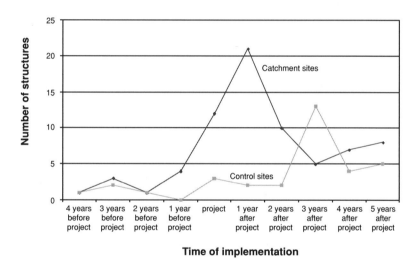

Figure 6.2 Implementing soil and water conservation (SWC) structures over time

In the farmers' opinion, the Catchment Approach was considered to be largely successful in Ojwando and Nyahera (both high-potential sites) due to the prevention of soil erosion on those farms where structures were implemented and the income earned from Napier grass. The project was considered to be successful to some extent in Wang'Adonji (low-potential site); but most farmers in Lower Nyabola (low-potential area) reported that the committee had been weak, not many people learned about the benefits of SWC and few farmers implemented the structures. For these reasons, few farmers in Lower Nyabola considered the Catchment Approach to have been successful. Similarly, in the control sites, there was considerable implementation of SWC structures in the high-potential control site, but much less in the low-potential control sites. The lower uptake of SWC structures in low-potential sites is largely due to the farmers' relatively low dependence upon crop agriculture as a form of income.

The popularity of Napier strips in the high-potential areas and the relationship to dairy farming have already been mentioned. In both Ojwando and Nyahera, for example, the Livestock Development Project had earlier provided a dairy cow to local groups and this provided an incentive to grow Napier grass as a fodder crop. Terraces (*fanya juu*) were implemented by only a small number of farmers because they were viewed as more labour-intensive and expensive, compared with Napier strips. Farmers also cited a lack of tools for the poor implementation of terraces. In the Nyabola control site, stone lines were established by two sample farmers because of the availability of stones in the area.

In the low-potential sites, sisal strips, aloe strips and unploughed strips were the most commonly used technologies. These are all relatively easy to establish and economical to maintain, particularly in view of the flooding that often occurs in the low-potential areas. Sisal leaves can be processed to provide fibre that is used for making ropes, mats and baskets. In many of the low-potential areas, the nature of the soil prevents the implementation of certain structures, such as *fanya juu* terraces and cut-off drains. In some areas, the hard pan (bed rock) in these soils is near the surface and constructing these structures is not easy using simple hoes. In other areas, deep unstable clay soils are so vulnerable to erosion that construction and maintenance of the structures is difficult. More importantly, the volume of water flowing from the watershed into the catchment is so high that these structures are quickly filled by soil sediment. Stone lines were implemented in hilly areas such as Wang'Adonji, where stones are readily available. However, farmers complained that livestock displace the stones. Similarly, the few farmers who planted Napier grass in Wang'Adonji reported that livestock and dry weather destroyed it.

Levels of technology adoption

With such a wide range of technologies implemented by farmers, it is difficult to present a concise analysis of adoption patterns. An index of adoption was therefore constructed in which levels of adoption were classified into three categories – high, medium and nil – the categories being defined according to the total number of structures on the plot visited, the total length of each of the three main categories of structures (strips, terraces and ditches, and other structures) and the level of maintenance (for those structures implemented subsequent to the project). Levels of adoption were calculated both for pre-project adoption (i.e. based on those structures that were first implemented prior to the project)[17] and also for post-project adoption (based on those structures that were first implemented during or after the project).[18] Examples from our sample farmers illustrating these various levels of adoption are given in Box 6.1.

Box 6.1 Farmers' levels of adoption

Haggai Odero Ogwai is a former committee member of Nyahera catchment in the high-potential area. He was born in 1936, educated in Kenya and Uganda, joined an auditing firm and later established his own construction company. He returned to Nyahera in 1982 as a farmer. Although he claimed to have maintained the terraces that had been constructed by his father during the colonial period, these were not evident at the time of the farm visit undertaken for the survey. The structures observed on the plot included a total of approximately 100m of Napier grass strips and 100m of unploughed strips. Two grass strips were originally established during 1996/1997 (the time that the catchment project was implemented in Nyahera); another grass strip was added in 1999; one unploughed strip was established in 1998 and a second in 1999. The level of maintenance of both types of structures was considered to be good. Given that there was no evidence of structures established before the project, Haggai was classified as a nil adopter prior to the project. Because the total length of strips did not exceed 300m, and the level of maintenance was not excellent, he was classified as a medium adopter post-project.

Dina Ondayi Okoth is a former member of the Wang'Adonji catchment committee in the low-potential area. Born in 1949, she attended primary school and later married and moved to Wang'Adonji in 1966 where she now lives as a widow. Prior to the catchment project, she was a contact farmer for the Ministry of Agriculture and had been trained in modern farming methods. She also recalls the colonial efforts to promote soil and water conservation (SWC). At the time of our visit, her farm had six unploughed strips (total 200m), one Napier strip (35m) and one *fanya juu* (35m). These were established in 1997 and 1998 and are maintained in excellent condition. Although the total length of these structures is less than 300m, Dina was classified as a high adopter for the post-project period due to the high level of maintenance of the structures. For the pre-project period, she was considered to be a nil adopter.

For the most popular structures (grass and unploughed strips) implemented by post-project adopters – including both medium and high adopters (see note 18) – the degree of adoption at the farm level was measured according to the density of strips per farm (measured in metres per hectare), the distance between structures (of any type) and the number of new strips constructed after the project (see Tables 6.7a and 6.7b). With the exception of farms with almost flat land in high- and low-potential areas, there has been a general increase in the overall density of strips constructed after the project. Similarly, the distances between structures pre- and post-project have decreased, indicating an increase in the density of SWC structures over time. Although one might expect the distance between structures to be shorter on the steeper slopes, this is not borne out by the data presented below. In general, the tables also show that there has been greater adoption in the high-potential areas: with the exception of figures for the distance between structures on farms with steep slopes, there is a higher density of strips constructed post-project in the high-potential areas, and the average number of new grass or unploughed strips is also greater. Yet, the distances between structures are relatively great (54m in the high-potential areas and 57m in the low-potential areas, compared with the recommended 15m mentioned above for a slope of 27 per cent). For the 31 cases where the existing records indicate the extension agent's recommendation for a particular farm, only six farmers actually achieved this and, on average, distances were almost twice (1.7 times) what the plan laid out. Although the increased distances between the structures could perhaps be regarded as an innovation on the part of the farmers, it seems likely that various constraints limited their abilities to establish the recommended level of conservation structures.

Table 6.7a *Number and density of grass/unploughed strips in high-potential sites*

		Almost flat slope: less than 15 per cent (n = 2)	Gentle slope: 15–20 per cent (n = 15)	Steep slope: 15–20 per cent (n = 12)	Very steep slope: more than 30 per cent (n = 11)	Total (n = 40)
Density of strips per farm constructed pre-project (metres per hectare)	**Mean**	128	42	66	37	52
	Minimum	0	0	0	0	0
	Maximum	256	312	448	210	448
Density of strips per farm constructed post-project (metres per hectare)	**Mean**	36	87	106	49	80
	Minimum	0	0	0	0	0
	Maximum	72	264	346	199	346
Distance between structures pre-project (metres)[a]	**Mean**	107	117	168	126	134
	Minimum	21	19	25	33	19
	Maximum	194	462	525	364	525
Distance in between structures post-project (metres)[b]	**Mean**	43	54	51	64	55
	Minimum	21	13	5	25	5
	Maximum	65	167	131	122	167
Number of new strips per farm[c]	**Mean**	2	2	3	3.6	2.8
	Minimum	2	0	0	1	0
	Maximum	2	4	7	8	8

Notes: As for Table 6.7b.

Table 6.7b *Number and density of grass/unploughed strips in low-potential sites*

		Almost flat slope: less than 15 per cent (n = 13)	Gentle slope: 15–20 per cent (n = 8)	Steep slope: 20–30 per cent (n = 7)	Very steep slope: more than 30 per cent (n = 3)	Total (n = 31)
Density of strips per farm constructed pre-project (metres per hectare)	**Mean**	**82**	**9**	**138**	**0**	**68**
	Minimum	0	0	0	0	0
	Maximum	500	72	912	0	912
Density of strips per farm constructed post-project (metres per hectare)	**Mean**	**32**	**92**	**192**	**16**	**82**
	Minimum	0	0	0	0	0
	Maximum	165	336	888	48	888
Distance in between structures pre-project (metres) [a]	**Mean**	**73**	**96**	**53**	**169**	**84**
	Minimum	20	42	13	83	13
	Maximum	210	188	119	300	300
Distance in between structures post-project (metres) [b]	**Mean**	**67**	**71**	**21**	**70**	**58**
	Minimum	19	12	10	42	10
	Maximum	210	187.5	37	125	210
Number of new strips per farm [c]	**Mean**	**0.2**	**1.0**	**2.6**	**1.0**	**1.0**
	Minimum	0	0	0	0	0
Maximum		1	5	9	3	9

Notes: Slope angles were estimated by the field assistants and included estimates for both the gentlest part of the slope as well as the steepest. 'Almost flat' had slopes of less than 15 per cent; 'gentle slopes' were generally up to 20 per cent but greater than 15 per cent; 'steep slopes' were roughly 20–30 per cent; and 'very steep' was considered to be greater than 30 per cent.

[a] Calculated based on all structures constructed across the contour (terraces, ditches, vegetative strips, unploughed strips and stone lines). Pre-project structures include only those still visible at the time of our survey. Where there were no structures or only a boundary hedge, the distance was taken to be the total length of the plot.

[b] Calculated based on all structures constructed across the contour (terraces, ditches, vegetative strips, unploughed strips and stone lines).

[c] Only grass strips and unploughed strips are included here.

Tables 6.8a and 6.8b illustrate the levels of post-project adoption for different types of sample farmer. As with Tables 6.7a and 6.7b, these tables show generally higher rates of uptake of SWC technologies in the high-potential villages. In the high-potential areas, the catchment villages show somewhat higher rates of change than the control sites, while this difference is less clear in the low-potential areas.

Table 6.8a *Levels of adoption in high-potential sites (percentage of each farmer type)*

Adoption class	Committee members (n = 10)	Farmers with laid-out farms (n = 17)	Farmers with non-laid-out farms (n = 9)	Control sites (n = 18)
No adoption	10	0	44	24
Medium adoption	60	65	44	54
High adoption	30	35	11	22

Table 6.8b *Levels of adoption in low-potential site (percentage of each farmer type)*

Adoption class	Committee members (n = 11)	Farmers with laid-out farms (n = 16)	Farmers with non-laid-out farms (n = 9)	Control sites (n = 38)
No adoption	46	75	44	58
Medium adoption	18	19	44	37
High adoption	36	6	11	5

The wide range of technologies makes it challenging to draw firm conclusions regarding the type of farmer who is likely to invest in SWC activities; but Table 6.9 presents some simple comparisons that indicate factors which may help to explain differences in behaviour. It compares those farmers who did not make any changes in their farms during or after the catchment projects (post-project nil adopters) with those who made at least one modification (post-project medium and high adopters – see note 18).

Table 6.9 indicates some differences between post-project adopters and non-adopters. In terms of agricultural resources, adopters tend to have larger holdings. Although some of the technologies (e.g. *fanya juu* or other terraces) are most relevant for steep fields, others (e.g. vegetative strips) are more appropriate for gentle slopes. Since the level of adoption was determined by the extent of implementation by either/or of these technology types, adoption is not concentrated on the steepest land. In the high-potential areas (where most adoption is found), adoption is much more likely among farmers for whom crop sales are an important source of income, and much less likely for households that depend upon business or petty trade. A series of logit analyses (not reported here) confirmed most of the significant relationships evident in Table 6.9. In high-potential areas, those households that depended more upon business or petty trade income were less likely to take up SWC technologies and those with higher crop incomes were more likely to adopt; larger landholdings were also significantly associated with adoption. In the low-potential areas, business and petty trade were again associated with lower adoption, and the ability to hire labour for weeding was related to higher uptake of SWC technologies.

Table 6.9 *Factors relating to the adoption of SWC technologies*

Characteristic	High-potential area			Low-potential area		
	Non-adopters (n = 13)	Adopters (n = 41)	Statistical significance	Non-adopters (n = 43)	Adopters (n = 31)	Statistical significance
Percentage farms with flat or gentle slopes	54	43	ns	73	68	ns
Percentage low labour availability	31	9	<0.1	43	32	ns
Percentage crop sale as most important source of income	15	59	<0.01	19	13	ns
Percentage business or petty trade as most important source of income	39	7	<0.01	40	23	ns
Percentage income from casual labour	14	12	ns	13	7	ns
Number of organizations	1.1	1.5	ns	1.3	1.5	ns
Landholding (acres)	3.2	6.8	ns	2.8	3.6	ns
Percentage female	39	37	ns	44	42	ns
Percentage hire labour for weeding	46	68	ns	53	70	ns
Number of months before buying grain	7.4	8.9	ns	4.7	4.8	ns
Number of cattle owned	2.0	3.0	ns	2.5	4.8	<0.05
Percentage with some education	69	82	ns	84	84	ns

Note: ns = not significant.

As Table 6.10 shows, there is a correlation between pre- and post-project levels of adoption. Fields that had SWC structures established before the project were more likely to establish further activities subsequently.

Table 6.10 *Adoption levels pre- and post-project*

Adoption level pre-project	Adoption level post-project		
	None	Medium	High
None	38 (68%)	28 (54%)	11 (55%)
Medium	17 (30%)	20 (39%)	5 (25%)
High	1 (2%)	4 (8%)	4 (20%)
Total	56 (100%)	52 (100%)	20 (100%)

Note: Pearson Chi-Square = 8.97; significance <0.1.

The quantitative data presented above are largely consistent with the more qualitative data collected. For the small number of project farmers who did not adopt any SWC, the reasons given for non-adoption included off-farm commitments; lack of labour and resources; or no planting materials for Napier strips. For the non-project farmers who did not implement SWC structures, reasons included the fact that they did not own land of their own; they were not active in farming (i.e. other off-farm commitments); poor health; no soil erosion on their farms; old age and lack of resources and labour; and lack of training from extension agents.

Farmers' sources of information about soil and water conservation (SWC)

The survey asked what the farmer's source of knowledge had been for each of the different technologies that had been implemented on the plot visited. The data show that the project was the principal source of information for the SWC technologies in the catchment villages (in both high- and low-potential sites).[19] This is particularly true for the major technologies emphasized by the project – for example, various types of strips and terraces/ditches. In the control sites, where SWC technology adoption was lower, farmers were a relatively more important source of information, particularly in the low-potential sites.[20] In the high-potential sites, non-governmental organizations (NGOs) were also an important source of information.

The data contain numerous claims of farmer-to-farmer learning, particularly in the control sites (see Box 6.2 and note 19). Without exception, every one of our key informant farmers reported having either taught or learned about SWC from other farmers; but it would be very misleading to assume that this sharing of knowledge necessarily led to the construction of SWC structures. Figure 6.2 suggests that there was a spread of knowledge from farmers in the catchment sites to farmers in the control sites, allowing some of the latter to implement the structures a few years after the project in the catchment sites. But the Catchment Approach was not the only source of SWC knowledge. In cases where farmers reported having copied particular structures from other farmers, they had obtained this knowledge either within their own village or further afield – often in their original home village (in the case of married women) or in other parts of the country where they had worked (in the case of men). In the high-potential sites, knowledge about grass strips was the most popular type of knowledge that was passed on to others. In the low-potential sites, knowledge about line planting and unploughed strips was most popular. It is important to note that structures themselves are very visible, allowing farmers to copy the things they see. Others reported having developed ideas for themselves, or learned from colonial agricultural officers, parents or grandparents, and NGO projects, clearly indicating that ideas about SWC have been around for some time. There were also one or two examples of misguided SWC efforts among non-project farmers: for example, leaving runoff routes unplanted (the farmer's own idea) and the incorrect use of trenches (dug along the slope, not the contour) that encourage runoff, with the idea of making the water flow out of the farm.

Box 6.2 An example of farmer-to-farmer learning

Gabriel Migele is a carpenter and a farmer in the Upper Konditi control site. He was born in 1933 and attended primary school up to class three. He farms 1 acre of land which he inherited from his parents. He first learned about soil and water conservation (SWC) when he was a boy from a white farmer known as Bwana Gimba, who had introduced cotton, rice and contour ploughing into the area. Gabriel's elder brothers started implementing contour ploughing on their farms and Gabriel learned more from them. Gabriel has never received any visit from the extension services; most of his agricultural knowledge comes from his relatives, neighbours and people for whom he builds houses. He also relies upon experiential learning. He maintains grass and sisal strips to prevent soil erosion on his farm. He learned about these structures from his neighbours in the Lower Nyabola catchment site, in particular Jashon Adongo (a retired extension officer and former catchment committee member) who is a great friend of his. Gabriel saw many people in Jashon's village implementing these technologies (often while he was roofing clients' houses, which gives him a clear view of what people are doing on their farms). He implemented the structures for himself and says that his own neighbours have also learned from him.

Experimentation, modifications and innovation

Only three farmers in the entire sample reported experimenting with novel SWC methods. In the first case (a former committee member), the farmer described having experimented to compare line planting using a rope line and *jembe* with an ox plough and concluded that the *jembe* was better. In the second case (also a former committee member), the farmer was implementing an experiment with the International Centre of Insect Physiology and Ecology (ICIPE) and the Ministry of Agriculture to examine the effects of Napier plots and *Desmodium* intercrops on *Striga* and pest control in maize and sorghum (the 'push–pull' technique; see Chapter 2). In the third case (a non-project farmer), the farmer reported using trial and error experimentation with Napier grass strips in combination with *fanya juu*.

Approximately two-thirds of project farmers (committee members and catchment farmers with laid-out farms) reported having implemented SWC structures according to the original recommendations made at the time of the SWC project. However, a comparison of the maps made by the extension officers showing the recommended structures with the maps of the farms at the time of the survey reveals that – for most of these farmers – not all of the recommended structures were actually constructed, some have not been maintained and some of the more labour-intensive structures (e.g. stone lines) have been replaced by less labour-intensive structures (e.g. grass strips).

Where innovations in the design (as opposed to the location, length or number) of specific structures were noted – as reported by about one third of project farmers – this tended to be through the combination of two or more structures – for example, the combination of trash lines, Napier, sisal or euphorbia strips with unploughed strips, terraces, stone lines or *fanya juu*. In Lower Nyabola catchment, the former catchment committee chairman (a retired extension officer) described how Napier grass strips were replaced by aloe and sisal strips after local farmers discovered that Napier could not survive drought and was eaten by free-grazing livestock. Elsewhere, one farmer reported having increased the width of an unploughed strip and allowing self-seeding trees and shrubs to grow on it for better runoff control. Another farmer reported increasing the width of a Napier grass strip in order to produce more livestock feed. Another reported adding aloe plants to sections of stone lines that had been damaged by runoff. In another case, a farmer had used his ox plough to construct a *fanya chini*. Most of these modifications appeared to stem from the farmers' own ideas, with some being copied from others. In one case, a female farmer said that she had borrowed a book about SWC from a neighbour.

Enhancing human capital: Increased knowledge

The Catchment Approach clearly enhanced the knowledge of farmers in the catchment sites, as reported earlier in relation to the sources of knowledge for the different SWC structures, and, to some extent, in the control sites through farmer-to-farmer knowledge transfers. The survey also asked about the importance of SWC and who should take responsibility for it, as well as seeking farmers' views about low external-input agriculture compared with high external-input agriculture (with particular reference to organic manure and chemical fertilizers). Most farmers believe that the importance of SWC relates to the prevention of soil erosion and the maintenance of soil fertility. It is largely considered to be the responsibility of individual farmers as opposed to a collective or community responsibility to conserve the soil, and that farmers and the government should work together to ensure soil conservation. These findings illustrate the relatively high level of environmental awareness in Kenya, which is due not only to the NSWCP but also to various other governmental and NGO activities, particularly since the early 1970s.

A detailed study of soil fertility replenishment systems conducted in western Kenya similarly found that farmers had uniformly high levels of knowledge about these systems (Place et al, 2003). Farmers in our sample were able to articulate clearly the advantages and disadvantages of external inputs such as fertilizer and manure, and there was no noticeable difference in the types of knowledge between project and non-project farmers – that is, they expressed largely similar views. However, in comparing the breadth of knowledge – expressed by both individual farmers and the sample as a whole – the non-project farmers were perhaps less articulate.[21] Whether this reflects a lower level of knowledge *per se* or a lower ability to express that knowledge is difficult to determine. However, given that fewer non-project farmers actually use fertilizer, it seems likely that their limited experience with fertilizer accounts for their apparently lower levels of knowledge.

An examination of the increase in knowledge over time also confirms the influence of activities other than the Catchment Approach that stretch back to the colonial period. Data from farmers in the catchment sites reveal that the largest proportion of new knowledge was gained during the catchment implementation period and its aftermath (1997–2003) and that this knowledge was gained within the village, through the extension staff. Overall (for all time periods), 80 per cent of farmers in the catchment sites said that they gained new knowledge or ideas from other farmers, and 85 per cent gained new knowledge or ideas from extension staff, school or other 'external' sources. Data from farmers in non-catchment sites, on the other hand, show that the largest proportion of new knowledge was gained during the period 1974–1988 (the early phase of the NSWCP) from outside the location, largely through other farmers and the Ministry of Agriculture. Overall, 90 per cent of farmers in the control sites gained new knowledge or ideas from other farmers, and 58 per cent from extension staff, school and other 'external' sources. Clearly, much knowledge was gained as a result of the project; but it remains difficult to quantify the farmer-to-farmer knowledge flows that originated from it.

Enhancing social capital: Local groups and organizations

Following implementation of the project, the catchment committees became dormant in all but perhaps one of the sample catchment sites. Former committee members in both Lower Nyabola and Wang'Adonji said that they felt it was not possible to continue to work as committee members without pay. Only in Nyahera was the committee apparently not disbanded after the project ended. Information collected from different informants, however, is not entirely consistent, and it seems likely that this group was established by a few (not all) of the former committee members. It has reportedly attracted many members and meets once a month to discuss issues that affect local farming activities and to seek possible ways of solving problems.

In each of the catchment sites, informants were able to name between one and four local groups whose establishment was reported to be in some way related to the catchment project. Most of these groups are involved in activities such as tree nurseries, dairy farming and – in one case – bee-keeping, which was initiated in the woodlots created with seedlings from the tree nurseries. Though related to SWC activities, they are not necessarily promoting SWC directly, and some receive outside support. Tree nurseries, for example, were promoted by the Forestry Department staff who worked in sites where the Catchment Approach had been implemented. In one case, the nursery specializes in fruit tree seedlings, and various fruits such as mangoes, oranges, lemons and avocados are now a common sight in the local villages. Tree nurseries tend to be managed by groups of local women who sell tree seedlings to local farmers. Several of these women's groups also provide credit to their members on a 'merry-go-round' basis, using the capital generated by the sale of seedlings. As mentioned earlier, dairy farming has been promoted by various projects through the introduction of exotic cattle and the promotion of zero grazing. The abundance of Napier grass in the high-potential areas has reportedly helped to promote dairy cattle among those who can afford them.

One of the local groups reported to have emerged as a result of the Wang'Adonji catchment project was the local branch of the Kiye Community-Based Development (KICOBADE) project. This is a district-wide community-based organization in Rachuonyo that takes care of orphans and offers counselling services to widows and those affected by HIV/AIDS. KICOBADE had already been established by the time of the Catchment Approach. Whether or not the local branch actually emerged as a result of the Catchment Approach is debatable; however, what is clear is that one of the former catchment committee members was instrumental in establishing the local branch. Given that committee members were originally chosen for their leadership qualities and because they displayed above-average networking abilities, it seems more likely that this particular organization would have emerged because of the abilities of the individual involved, which may well have been enhanced through her involvement with the catchment committee.

While it is difficult to determine precisely whether specific groups emerged as a direct result of the catchment project, the survey findings show clearly that there is greater membership of local organizations in the former catchment sites compared with the control sites: 82 per cent of households in the catchment sites claimed to have at least one member who belonged to one or more local organizations, whereas the figure was 63 per cent for the control sites. Within the catchment sites, families of former committee members tended to have the highest rate of membership: 91 per cent of the households of former committee members were represented in one or more organizations, compared with 78 per cent of the households of non-committee members. Group members in the catchment sites reported being more active than those in the control sites, and committee members tended to be by far the most active. Although we collected some information regarding the time of establishment of the main groups within each site, many more groups were mentioned by individual survey respondents, and we did not collect data on the membership levels prior to the project. It is therefore difficult to compare membership before and after the project. Women's groups were by far the most popular type of group throughout the survey area, followed by social welfare (self-help) groups and church groups. Interestingly, only three individuals in our sample reported belonging to agricultural groups, and no agricultural groups were reported in the focus group discussions. This is partly because other types of groups (notably women's groups) undertake agricultural activities; but it also suggests that agriculture *per se* is not a high priority for collective initiatives.

Summary

A wide range of different technologies were promoted by the project and adopted by farmers, with a clear preference for specific technologies in different areas – for example, Napier grass strips in high-potential areas and sisal or aloe strips in low-potential areas. There were considerably higher levels of adoption in the high-potential areas where farmers rely more upon crop agriculture as a source of income. The markedly lower levels of adoption in the control sites suggest that the farmer-to-farmer spread of technologies as a result of the project has been limited. Moreover, even where farmers in the control sites had adopted SWC technologies, this was not necessarily because of the catchment project. SWC information has been around for a long time and is promoted by various organizations. In general, those who had adopted SWC technologies prior to the project were more likely to adopt in the post-project period. Although adoption levels went up at the time of the project, the implementation of structures has continued after the project itself, possibly because farmers had seen positive results on either their own or other people's farms and chose to implement, or possibly in order to spread the labour inputs required for SWC technology construction over various seasons. In terms of farmers' development of the technologies themselves, there is little evidence of innovation. Where modifications took place, these tended to be either increasing the size of structures, or combining two different structures.

In terms of human capital, much knowledge was gained as a result of the project; but levels of environmental awareness are generally quite high anyway in the study area. This illustrates the limitations of a one-off project-based approach and emphasizes the importance of the cumulative effects of repeated or continued learning opportunities or awareness-raising (also illustrated by the relationship between pre- and post-project adoption). Although it is difficult to document farmer-to-farmer knowledge flows originating out of the project itself on a wide scale, a few examples of specific instances show that this certainly did take place. A comparison of uptake of SWC technologies in the different sample sites over time also suggests that there was some increase in human capital in the control sites as a result of farmer-to-farmer knowledge transfers emanating from the project some two years after its implementation.

In terms of social capital, the higher levels of membership of local groups and organizations within the catchment sites suggest that social capital is higher in these sites. Although this may have been influenced by the project, it is also possible that high levels of membership existed before the project and that this partly accounted for the relative success of the project in these sites. In most cases, the former catchment committees were no longer functional. Although other organizations and activities loosely related to SWC were established following the project, these were not specifically promoting SWC *per se*. At the individual level, it is important to note that those farmers most closely involved in the catchment project (the committee members) were those who already had high levels of social capital to begin with and that this was considerably enhanced in some cases (D. Nyantika, pers comm, 2004). Several of the committee members remarked that after the end of the project there was no further contact with the extension agents and that they would have liked some kind of 'reminder' from time to time about SWC. It is perhaps paradoxical that, although farmers' networks and organizations are expected to emerge and be sustained after LEIT projects, no effort was made to sustain the networks established between extension agents and farmers by the project itself. This again illustrates the limitations of project-based programming in which activities are relatively short-lived. Indeed, the one-year time frame of the Catchment Approach was regarded as too short, and although it had been hoped to work for longer in each area under the current National Agriculture and Livestock Extension Programme, this has unfortunately not been possible.

Annex 1: Methodology

Data were collected in three phases:

1   background information and site selection;
2   initial fieldwork;
3   follow-up fieldwork.

In the first phase, written documents were reviewed and key informants[22] were interviewed at national and regional levels to gain an understanding of the project as a whole and how it was implemented in western Kenya. Ten divisions were initially selected through discussions with agricultural staff and those familiar with the area as containing both high- and low-potential sites, cultivated to a variety of crops (that is, excluding fishing communities and sugar-producing areas), where NSWCP was well implemented, and where few other agencies or projects have since been operating. Division offices were visited, and the existing records[23] from the catchments implemented between 1994/1995 and 1998/1999 (inclusive) were noted.

A database containing brief details on each of the 128 catchments implemented within these divisions during the period was compiled and used to draw up a shortlist of ten catchments in high- and low-potential areas that involved a range of different SWC structures and that appeared to have been successful (a good number of farms laid out and implemented), but not outstandingly so (e.g. having won national prizes). Reconnaissance visits to each of these ten catchments allowed for the selection of four case study catchments (two high-potential and two low-potential areas). At the divisional and local levels, the research team held personal informal interviews with divisional soil conservation officers, catchment committee chairmen and other opinion leaders. Three control sites (where the NSWCP had not been implemented) were also selected. The control communities had characteristics roughly similar to those of the selected catchments but without intervention by the NSWCP. They were used to judge the degree of agricultural change and innovation without project intervention and to measure the degree of spontaneous diffusion to non-project communities.

In the initial fieldwork phase (approximately four days per catchment site), meetings were held with the divisional agricultural staff and the members of the former catchment committee. Existing project records were consulted and sample farmers for the questionnaire selected (see below). Farm visits were made; a small number of questionnaires were used with the committee members, and focus group discussions and other PRA exercises were undertaken. Control sites were selected. In the follow-up field visit (approximately three days per site), the remaining questionnaires were completed and key informant interviews with farmers took place. Thus, in each study site, four main methods were used for the collection of qualitative and quantitative data: farm visits; focus group discussions and other PRA exercises; a questionnaire survey; and detailed interviews with key farmer informants. Much of the fieldwork was undertaken with a staff member from the division agricultural office who provided additional information.

Purposive sampling was used to identify farmers for the survey: in each of the catchment sites it was originally planned to include six committee members and six non-committee members whose farms were laid out (e.g. 12 'project' farmers), plus six non-committee members whose farms were not laid out or who were not present at the time of the project (non-project farmers). This sampling method was implemented as far as possible. The names of all the committee members in each catchment had been recorded in the project records, which also listed all the farmers who were present at the time of the project's implementation. The names on these lists were checked and used to draw up a list of those who were still living in the community and whose farms were laid out, as well as a second list containing the names of the farmers whose farms were not laid out, plus the names of the farmers who started farming in the area *after* the project (for example, new households that had emerged or farmers who had immigrated or returned to the area). In Lower Nyabola and Ojwando it proved difficult to find enough non-project farmers, and in some cases sampled farmers were subsequently re-categorized, based on the answers they gave to the questions. For these reasons, there was some deviation from the planned sample sizes for each categorization (see Table 6.4). In non-project sites, there were no lists of households existing in the village, so the researchers sat with the village elders to compile a complete list.

Two farmers per site were also selected as key informant farmers (from among those who had been included in the survey). These farmers were identified, based on the contacts made during the first phase of the fieldwork, as those who were practising soil and water conservation and (in the sites where the project was implemented) were knowledgeable about the Catchment Approach. The key informants included at least one catchment committee member and one non-committee member. All key informant farmers had good social networking skills (i.e. were leaders of other local community groups) and had the ability to express themselves clearly.

Notes

1   We are grateful to David Nyantika, Brent Swallow, Qureish Noordin, Samuel Muriithi and Frank Place of the World Agroforestry Centre (ICRAF) for their guidance in designing the research, selecting the field sites and reviewing the final report upon which this chapter is based. Particular thanks go to the farmers in the study sites who contributed their time to answer our questions and show us their farms, and to the various divisional agricultural staff who assisted with the fieldwork. Mr Mukono of ICRISAT diligently entered all the quantitative data into the databases. Invaluable assistance in the statistical analysis was kindly provided by Eliud Lelerai and Richard Coe (ICRAF). The map in Figure 6.1 was provided by Meshack Nyabenge (GIS Unit, ICRAF). We are grateful to the International Livestock Research Institute (ILRI) for allowing Nelson Mango the opportunity to complete the project. Colin Poulton provided detailed comments on an earlier draft, and Japheth Kiara and his colleagues at NALEP offered useful feedback

2   The branch had three support sections at headquarters – namely land husbandry (soil and agroforestry), water management (water harvesting and water conservation), and planning and education (monitoring and evaluation, socio-economics and training).

3   As illustrated by the creation of the Permanent Presidential Commission on Soil Conservation and Afforestation. The commission was charged with the responsibility of coordinating all institutions involved in the conservation effort.

4   Additional activities included mobilizing farmers when collective tasks were to be performed; taking care of tools provided by the project; initiating and maintaining a tree nursery as well as bulking plots; collaborating with extension officers in taking farm measurements; distributing items provided by the project; meeting and ushering in visitors to the catchment; encouraging schools to assist in soil and water conservation; ensuring that farmers established structures on their farms according to the instructions of the extension officers; ensuring that established structures were protected and well maintained (ensuring that cattle did not destroy already established structures); and encouraging farmers to adopt modern farming practices.

5   In our sampling process, we purposely avoided areas where such cash crops are predominant in order to avoid extreme differences in cropping patterns between the sample sites in each zone.

6   For the 16 households which cultivated sorghum in the high-potential areas, average annual sorghum production was 186kg.

7   Among our sample farmers in the high-potential zone, 19 per cent did not own any cattle, 70 per cent owned one to four cattle, and 11 per cent owned five or more cattle.

8   27 per cent of sample farmers in the low-potential area did not own any cattle, 45 per cent owned one to four cattle, and 28 per cent owned five or more cattle. Cattle in the low-potential areas tend to be the local zebu type.

9   Of the 18 non-project farmers in the catchment sites, 11 had been present in the catchment at the time of the project but did not have their farms laid out, and 7 were not present in the catchment at the time of the project. The sampled farmers in each of the catchment sites included both types of non-project farmer.

10   It is not clear from the project records how 'implementation' was defined in this case.

11   Amount of household labour was determined according to a scoring system for the types of labourers available for the two main planting seasons. Points for the long rains growing season were allocated as follows: full-time worker – 10 points;

part-time worker – 5 points; weekend worker – 2 points; irregular worker – 1 point. Half the number of points was allocated for each of these categories for the short rains growing season. The total number of points was then used to determine the following categories: high labour availability – 32 or more points; medium labour availability – 15 to 32 points; low labour availability – 14 points or less.

12   Based on the findings of PRA exercises undertaken in the field sites, levels of wealth for the quantitative data set were determined according to house type (wealthy people's houses were those with tiled or iron sheet roofs; poor people's houses were those with mud walls); cattle ownership (wealthy people had at least three cattle; poor people had none); size of landholding (wealthy people had holdings of more than 5 acres; poor people had less than 2 acres); and whether or not labourers were employed for weeding (wealthy people employed labourers for weeding; poor people did not).

13   Fields with steep slopes were generally more than 30 per cent and ranged up to 90 per cent.

14   The NSWCP began in 1974. Since there were very few structures established during the immediate post-independence period, the periods of 1963–1973 (the 'Lost Years') and 1974–1997/1998 (the NSWCP prior to the implementation of the Catchment Approach in the case study sites) have been combined in the table.

15   The Ministry of Livestock Development implemented the National Dairy Development Project (NDDP) in specific districts from 1980 to 1998 with the support of the Netherlands government. The NDDP promoted the concept of zero grazing through a package of various components, including the production of Napier grass as a high-yielding fodder crop. When the NDDP ended in 1998, the same technical package was promoted by the Finnish-funded Livestock Development Project (LDP) – a relatively small project focusing on women's groups in western Kenya.

16   The time of project implementation for the control sites is defined according to the neighbouring sample catchment site.

17   Nil adopters were those who did not implement any SWC structures prior to the project. Medium adopters were those who implemented up to 300m of either strips (unploughed, grass, sisal or aloe), terraces/ditches (*fanya juu, fanya chini,* retention ditch, trench, check dam or terrace) or other structures (banana lines, hedges, tree lines, trash lines or stone lines). High adopters were those who implemented more than 300m of either strips, terraces/ditches or other structures.

18   Nil adopters were those who did not implement any SWC structures subsequent to the project. Medium adopters were those who implemented up to 300m of either strips, terraces/ditches or other structures, and low or medium levels of maintenance overall. High adopters were those who implemented more than 300m of either strips, terraces/ditches or other structures, or who had fewer structures but with high levels of overall maintenance.

19   Among farmers in the high-potential catchment sites, knowledge of about 61 per cent of the structures had been gained from the catchment project and 30 per cent from other farmers; the remaining 9 per cent came from 'other' sources. In the low-potential catchment sites, the respective figures were 67 per cent, 24 per cent and 9 per cent.

20   In the high-potential control sites, knowledge of about 36 per cent of SWC structures came from other farmers, 36 per cent from 'other' sources and 28 per cent from the Ministry of Agriculture. In the low-potential control sites, the respective figures were 65 per cent, 29 per cent and 6 per cent. The major example of 'other' sources was NGO projects.

21   For example, 87 per cent of project farmers remarked on the high cost of fertil-
      izer, whereas 59 per cent of non-project farmers mentioned this point; 48 per cent
      of project farmers and 32 per cent of non-project farmers commented on the long
      residual effect of manure; 43 per cent of project farmers and 25 per cent of non-
      project farmers said that organic manure is bulky and labour-intensive; 37 per
      cent of project farmers and 25 per cent of non-project farmers said that organic
      manure retains moisture in the soil; but 24 per cent of project farmers and 30 per
      cent of non-project farmers mentioned that fertilizer does not do well because of
      low rainfall.

22   Personal interviews with national and local authorities, project experts and donor
      representatives individually and in groups, including interviews with selected
      senior government officials from the SWC branch of the Ministry of Agriculture
      in Nairobi, expatriates from Sida in Kenya, a scientist from ICRAF, and an offi-
      cer from the Permanent Presidential Commission for Soil Conservation and
      Afforestation in Nairobi.

23   The level of documentation available for the NSWCP is particularly impressive,
      and we wanted to be able to draw on the existing records for the research. For
      each catchment, such records included a catchment map; a black notebook
      containing sketch maps and details of each individual farm; PRA reports; names
      of the committee members; and records of committee meeting minutes, training
      visits and other correspondence.

7

After School: The Outcome of Farmer Field Schools in Southern Sri Lanka[1]

Robert Tripp, Mahinda Wijeratne and V. Hiroshini Piyadasa

Introduction

The farmer field school (FFS) is a well-established technique for introducing principles and methods for crop management through hands-on learning. It has had its most widespread application in promoting integrated pest management (IPM) in Asian rice systems. This chapter describes the experiences of a FFS programme focused on IPM and other crop management techniques for rice farmers in Sri Lanka. It is particularly concerned with exploring the relationships between farmers' human and social capital, on the one hand, and the innovations presented in the FFS, on the other. The study addresses practical concerns for understanding what type of farmer is likely to take advantage of a FFS, what type of knowledge is gained through participation, and whether this knowledge is likely to spread to other farmers or contribute to capacities for further innovation.

After an introduction to rice cultivation and IPM in Sri Lanka, the next section presents a description of the research area and the organization of the study. This is followed by sections that discuss the type of farmer who participated in the FFS; the principal outcomes related to practices, knowledge and farmer innovation; the diffusion of information from FFS farmers to others; and the degree to which farmers' access to information and labour is likely to affect the utilization of the practices recommended in the FFS.

Rice, extension and integrated pest management (IPM)

Rice in Sri Lanka

Rice is by far the most important food crop in Sri Lanka. Average per capita consumption is about 100kg per year, providing 45 per cent of the calories and 40 per cent of the protein in the diet of the average Sri Lankan. In the past, Sri Lanka has often had to import a significant amount of rice; but recent government policies encourage self-sufficiency, and production in 2002 satisfied about 90 per cent of domestic requirements. Rice is grown in most parts of the country, under diverse conditions, including major irrigation schemes, minor irrigation structures and some rain-fed cultivation.

Rice is grown during two seasons: *Maha*, corresponding to the north-east monsoon from October to February, and *Yala*, corresponding to the less reliable south-west monsoon from April to July. Approximately 560,000ha of rice are grown during *Maha* and 310,000ha during *Yala*. Most rice cultivation relies on fairly high levels of external inputs, and although insecticides have never been subsidized in Sri Lanka, their easy availability and convenience have led to increasing use. The problem of high insecticide use has been sufficiently serious to attract extension attention for the past two decades.

Extension strategies

Agricultural extension has passed through various phases in Sri Lanka during recent years. The Training and Visit (T&V) system was introduced through a World Bank-funded Agricultural Extension and Adaptive Research Project (AEARP) in 1979. The T&V system reoriented Department of Agriculture (DoA) extension strategies; but it was not financially sustainable and had limited impact (Wijeratne, 1988). Some efforts were made to introduce improved pest control methods through T&V, but these had little success. A major government decentralization reform in 1990 removed village-level extension workers from centrally controlled DoA offices at district level and reassigned them to local government offices. They became multi-duty officers (*grama niladhari*) and there is evidence that this reform lowered their contact with farmers (Wijeratne, 1993). As a result of the reform, the DoA has fewer personnel available to carry out agricultural programmes and must increasingly rely upon its middle-level agricultural instructors (AIs). One of the major extension strategies in recent years has been the *Yaya* programme, in which AIs organize a group of farmers who meet several times during the season at a demonstration plot. The programme delivers messages on a set of priority crop management topics, including more rational pest control.

Farmer field schools (FFS)

The most important effort by extension to reduce insecticide use has been the FFS programme, which is the focus of this chapter. It has been part of a long-term commitment to IPM in Sri Lanka since the mid 1980s, led by the United Nations Food and Agriculture Organization (FAO) and funded by various donors. The principal FFS activity took place between 1995 and 2002, during which time more than 600 FFS were organized throughout the country. Most of them were run by AIs (or staff of the independent Mahaweli Authority responsible for the nation's major irrigation scheme) who had attended one of the four-month courses for facilitators offered in Sri Lanka under an FAO project. In later years, some of the FFS were run by farmer graduates who were given additional training and support by the DoA as 'community IPM'. This latter initiative tried to address the relatively high cost of FFS by transferring more responsibility for spreading the IPM message to trained farmers. Staff of several NGOs have also been trained by the DoA and have organized their own FFS.[2]

The farmer field schools in rice take place during the crop season. A typical FFS includes about 20 participants who meet one day a week, from early morning until midday, for a total of approximately 14 sessions. The FFS is distinct from the top-down approach of most extension activities, and participants are encouraged to question, contribute and exchange observations. The participants always meet in the same field. Early sessions introduce the techniques of agro-ecosystem analysis and during the course the farmers take part in various exercises (such as making collections of pests and natural enemies) to aid their understanding of plant–insect and pest–predator relationships. Simple experiments (such as leaf-cutting to assess the effect of leaf damage early in the crop cycle) are also conducted. The original focus of the FFS was preponderantly IPM, with farmers being introduced to a range of information that allows them to appreciate biological insect control, to question the need for insecticides and to consider alternative methods of insect control. But other subjects have gradually been introduced, and in each session the facilitator presents topics in crop management relevant to the particular growth stage of the crop. Facilitators estimate that these additional topics eventually constitute up to half the time of the FFS.

Research area and organization of the study

Research sites

The study examined the outcomes of FFS activities in Sri Lanka's Southern Province (see Figure 7.1), which features a range of rice farming conditions, including three climatic zones. Dry zone areas in the east have an annual rainfall of less than 1200mm and rice cultivation is mostly dependent upon large irrigation works; some of these farmers pursue dryland farming as well. The intermediate zone has somewhat higher rainfall and rice farming is often complemented with the cultivation of tree crops. The wet zone, occupying the western part of the province, receives rainfall in excess of 2000mm and features the smallest holdings, a higher dependence upon tree crops and the most urbanized conditions, associated with a high degree of off-farm labour. The study included sites in Hambantota (dry and intermediate zones) and Matara (wet zone) districts; the province's westernmost district (Galle) was not considered for the study because its highly urbanized character makes it less representative of major rice-growing conditions.

We imposed several conditions on the selection of sites for the study. In order to assist in assessing long-term consequences, we chose sites where a FFS had been conducted at least five years earlier. Sites were also chosen to represent the range of rice-growing conditions in the two districts. In addition, we only selected sites where we had evidence that the FFS had been well managed (and only where a qualified DoA facilitator had been in charge). We conducted interviews with the local extension administration and made preliminary site visits to ensure that the FFS examples included in the study had been competently conducted and well attended.

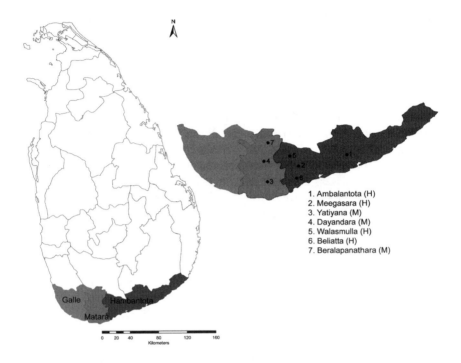

Figure 7.1 Locations of study sites in Sri Lanka

Although it was possible to meet these conditions for site selection, several factors limited our capacity to include the broadest possible range of examples. In the first place, the conduct of FFS depended upon the interest of local extension offices and access to trained FFS facilitators. These resources and commitments were not evenly distributed, and FFS activity tended to cluster in certain areas. In addition, a few areas in Hambantota that had been earlier FFS sites experienced severe drought in the seasons immediately prior to the study, sufficiently affecting the extent and nature of rice growing to make assessment of current practices problematic. Finally, we eliminated those candidates that were very close to urban areas where rice farming is of lesser importance (and where extension had more difficulty in attracting farmer interest in FFS). Nevertheless, we feel that our sample provides a reasonably broad picture of the outcome of well-managed FFS in major rice-growing areas of Southern Province. There were a total of 56 sites (34 in Matara and 22 in Hambantota) from which to choose; we reviewed most of these candidates at the desk level and then conducted field visits to about one third of them before selecting the final sample.

Organization of the study

The sample covers seven locations. In each location, three types of farmers are included. The sampling was based on *yayas*, or irrigation tracts. A *yaya* is a group of contiguous paddy fields dependent upon a single irrigation source. A village (*gama*) can include one or more *yayas*. First, we selected a *yaya* where a FFS had been conducted and randomly chose ten farmers from the list of graduates (roughly a 50 per cent sample). Second, we obtained a list of all farmers in the *yaya* and randomly chose ten non-participants (the *yayas* in our sample vary in size from 40 to over 200 farmers; hence, the sampling proportion varies accordingly). Third, we identified a *yaya* of a different nearby 'control' village that was as equivalent as possible to the focus *yaya*, particularly with respect to irrigation quality and farm size, and randomly selected ten farmers for interview (these control *yayas* were between 2km and 5km from the focus *yaya*). Thus, each location comprises 30 farmers (participants, neighbours and controls) and the entire sample totals 210 farmers.

A questionnaire was administered during the *Yala* 2003 season. The questionnaire was developed after informal discussions with farmers in various locations exploring a range of topics; several drafts were then tested and refined and three trained enumerators, recent graduates in agriculture from the University of Ruhuna, administered the questionnaire. Enumerators introduced the questionnaire to farmers as a study of rice production practices, rather than as an assessment of FFS, and the questionnaire explored various topics on rice production practices before examining issues related to participation in, or knowledge of, the FFS. The most detailed questions concerned practices in the past *Maha* season (2002–2003).

The quantitative data reported in this chapter are the results of the questionnaire. These are complemented by observations made during informal surveying of farmers in the study area as well as in other locations in southern Sri Lanka.

Farmer characteristics and practices

The results shown in Table 7.1 present averages for all the sample farmers in each location. The table illustrates the small-farm character of rice cultivation in Southern Province, with an average holding of about 0.9ha. Slightly more than half (52 per cent) of the plots were held under *ande* tenancy, in which tenants have secure access to their fields but provide one quarter of the harvest to the (mostly absentee) landowners.[3] 40 per cent of the farmers owned their plots and 8 per cent leased their fields for cash. Rice is an important source of livelihood in the study area, accounting for 43 per cent of household cash income (agricultural activities combined account for about two-thirds of household income). In all seven locations, farmers plant rice in both *Maha* and *Yala* seasons.

Despite the small holdings, 82 per cent of the survey households report that they are always self-sufficient in rice and less than 5 per cent report that they must buy some rice every season. This distinguishes the sample farmers from many rice growers in Sri Lanka who do not produce enough for household needs. Although self-sufficient rice cultivation is a cultural ideal in rural Sri Lanka, it is not often realized.[4] Because our study concentrates on important rice-growing areas where successful FFS have been conducted, the resulting sample comprises households which, despite limited resources, have a significant advantage over many other rural Sri Lankan households that do not have access to paddy land. The average paddy yields in the sample are 3.8 metric tonnes per hectare (or about 2.5 metric tonnes of milled grain per hectare) and are comparable with average rice yields reported for Sri Lanka (Central Bank of Sri Lanka, 2002). A little over half the farmers have access to land for tree crops; coconut, tea, cinnamon and banana are the most important. Only a small minority of the sample farmers have *chena* land for shifting cultivation where they grow crops such as finger millet, sorghum or sesame.

Table 7.1 *Research locations*

Location	Year of farmer field school (FFS)	Number of farmers in FFS *yaya*	Average paddy area (hectares)[a]	Average paddy yield (metric tonnes per hectare) [a]	Percentage cash income from paddy sale [a1]
Hambantota District					
1 Ambalantota	1994	92	1.40	6.10	65
2 Meegasara	1998	40[b]	1.08	4.16	56
3 Walasmula	1995	211	1.04	3.39	55
4 Beliatta	1996	90	1.00	3.73	36
Matara District					
5 Yatiyana	1997	150	0.88	3.10	42
6 Dayandara	1998	121	0.48	3.61	32
7 Beralapanathara	1997	48	0.36	2.19	15

Notes: [a] Average for all types of farmer in the location.
[b] Combined participants from two small yayas.

Although there is much similarity between the locations, a few differences stand out. Location 1 is a high-potential dry zone area, well connected to major roads, and features relatively larger holdings than average. It has been the focus of considerable extension attention (and is located near a major government rice research station). These factors help to explain the high yields observed in this location. In contrast, location 7 has the smallest holdings and lowest yields in the sample. This location relies heavily upon smallholder tea production and farmers devote more attention to this crop than to rice. In the following discussion, most analyses will include all seven locations; but there are certain instances where we eliminate locations 1 and 7 because of their unusual characteristics.

A comparison of the characteristics of the focus *yayas* and the control sites shows no important differences. The average rice holding and yields in the *yayas* where the FFS were conducted are slightly higher than in the controls; but the differences are not statistically significant. The relative importance of paddy as a source of cash income is equivalent for the two groups, as is the relative importance of non-agricultural income sources.

Most of the basic practices of rice production are common throughout the research area. Many of these practices represent a significant trend towards labour-saving technology. Hired labour is an important factor of production for most farmers; most of the labour is drawn from nearby households.[5] A minority of the labourers are themselves rice growers; most are landless or have only rain-fed plots. Most rice farmers use two-wheel tractors to prepare their fields; about 12 per cent of our respondents own these tractors, while the majority rent them. All of the farmers plant by broadcasting pre-germinated seed, rather than the more laborious transplanting that was previously the norm; 64 per cent of the respondents use hired labour for planting. Virtually all farmers use synthetic fertilizer in their rice fields, making three or four applications per season. Most weed control is with herbicides (86 per cent make a single application), but 10 per cent do some additional hand weeding. When insecticides or fungicides are used, most (70 per cent) of the farmers hire labour for spraying. The major exception is those farmers (about one half) who own a sprayer, half of whom report applying insecticides themselves. Harvesting and threshing are carried out with the help of hired labour in 94 per cent of cases.

For the 140 sample farmers who did not participate in a FFS, the average number of insecticide applications in *Maha* 2002–2003 was 1.7; 63 per cent of these farmers made at least one application. The insecticide management of these farmers is equivalent to behaviour noted in other studies of insecticide use in Sri Lanka. Nationwide data collected by the Department of Census and Statistics in 2000 indicate that roughly 70 per cent of rice fields received at least one application of insecticide. A study conducted for FAO to evaluate the FFS programme recorded an average of 2.2 insecticide applications for farmers who did not participate in a FFS (van den Berg et al, 2002). A study conducted in 1995 showed about 68 per cent of all farmers using insecticide on rice; those who used insecticide made an average of 1.8 applications (Nugaliyadde et al, 1997).

Who participates in farmer field schools (FFS)?

Selection of FFS sites

A farmer's opportunity to participate in a FFS depends upon local DoA decisions regarding eligible communities. Resources are available to organize only a relatively few FFS, and extension offices must select their targets. Extension staff say that their priorities for a FFS are based on need (high insecticide use or serious pest problems); but it was not possible to ascertain the degree to which these rules are followed. In addition, extension staff naturally tend to work in communities where they have good relations. In some cases these may be villages that have somewhat better resources; in several locations in our study there was a tendency to favour villages with access to a reliable irrigation source for a FFS. Each village in Sri Lanka is supposed to have a farmers' organization that serves as a link with government agricultural agencies. In practice, some of these organizations are quite active while others are moribund. In most cases, extension personnel contact the farmer organization in a village to see if there is interest. If so, they provide information about the nature of the FFS and interested farmers are enrolled. In the majority of cases, the FFS serves farmers who have plots in a single *yaya*; but members of several small, neighbouring *yayas* may be combined in one FFS.

FFS farmers and their neighbours

Even though activities such as FFS are theoretically open to all, there is the possibility that better-resourced farmers may take more advantage of them. In the case of the FFS in southern Sri Lanka, our data indicate that while there are some differences between FFS farmers and their neighbours, these probably do not represent any significant bias. Table 7.2 summarizes data on a number of points of comparison.

Table 7.2 *Farmer field school (FFS) farmers and their neighbours*

Characteristic	FFS farmers (n = 70)	Neighbours (n = 70)	Statistical significance
Total paddy area (hectares)	1.02	0.82	<0.1
Annual household paddy production (metric tonnes)	7.59	6.12	ns
Yield of focus paddy field (metric tonnes per hectare)	4.07	3.59	ns
Percentage cash income from paddy	44	42	ns
Percentage cash income from non-agricultural sources	33	38	ns
Percentage work as farm labour or casual labour	13	40	<0.001
Education of household head (years)	7.2	6.9	ns
Number of group memberships of household	1.9	1.8	ns

Note: ns = not significant.

The farmers who enrol in the FFS tend to have somewhat larger rice holdings than the non-participants. The annual rice production and the yields of participating farmers are higher as well, but not significantly so. There are no significant differences between the two groups in the relative importance of rice as a source of household income, nor in the role of non-agricultural sources of income. However, there is a strong tendency for households with members who work as agricultural or casual labour to be less likely to participate in a FFS. This difference is consistent in all the research locations (on the other hand, the relatively few respondents who also have salaried jobs or practise a trade are at least as likely to join the FFS). In the more urbanized areas in the west of the province (which were not included in our study), extension staff often complain about the problems of achieving sufficient attendance at FFS because of the many 'part-time farmers' who have permanent off-farm employment and very small paddy holdings. Less than 10 per cent of the FFS participants in our sample were women, in most cases substituting for a husband who did not have time to attend.

The standard of education in the sample villages is relatively high (average of about seven years) and there is no difference in schooling between participants and non-participants. It is difficult to assess a farmer's indigenous technical knowledge (ITK); but the questionnaire included items on other methods of pest control. More FFS participants (57 per cent) than non-participants (41 per cent) mentioned at least one local control method with which they were familiar; but further analysis indicates that this difference is accounted for by information on alternatives introduced in the FFS.

There are few obvious differences between types of farmer with respect to social capital. That is, membership in particular social groups does not seem to influence participation in the FFS. Most farmers say that the level of cooperation in their own *yaya* is average, although FFS graduates are somewhat more likely than their neighbours to say that cooperation is above average. The FFS participants are no more likely than the non-participants to be 'joiners', and both groups report an equivalent number of memberships in various organizations. The widest membership is enjoyed by the local farmer organizations, to which the majority of farmers belong; 94 per cent of the FFS participants reported membership and 79 per cent of their non-participant neighbours are members. We asked these farmers about their level of participation in the farmer organization, and there was no difference between the FFS participants and their neighbours in the proportions that reported active and less active participation.

The second most common membership reported by farmers is in death donation societies that collect periodic contributions and provide support for funeral expenses. Three other types of organization were mentioned by at least 10 per cent of households: village development societies, *Samurdhi* groups and *Sanasa* groups. Village development societies are formed by various government institutions. *Samurdhi* groups are formed for specific activities such as digging wells and are connected to the government programme of the same name that provides dry rations and other assistance to the poorest members of a community. *Sanasa* groups are organized by rural banks to provide credit. There was no difference in membership in any of these groups between FFS participants and their neighbours. Several other organizations were mentioned by small numbers of respondents; but there is no obvious correlation between membership and participation in a FFS.

There are no significant differences between FFS farmers and their neighbours in terms of attendance at village-level agricultural extension activities. However, a difference between FFS farmers and others is that the former are much more likely to have participated in visits outside the village related to agricultural subjects (such as field days or short courses); about one quarter of FFS farmers had participated in at least one such event, as opposed to one tenth of their neighbours. The majority of the visits listed (for both types of farmer) were subsequent to the year of the FFS. The survey questions were not sufficiently precise to determine if this was because the FFS served to connect these farmers to other extension activities, or whether these farmers had always been more eligible for participation in special extension events.

In summary, the FFS is an opportunity open to all farmers in those villages selected by the DoA. Those who enrol tend to have somewhat larger rice holdings; but the differences are not great and farmers with very small fields participate successfully. On the other hand, the FFS is much less likely to attract farmers who work as agricultural or casual labour, probably because of the demanding weekly schedule. There is little evidence that participation in the FFS requires above-average human or social capital.

The effects of the FFS

What changes in behaviour or knowledge can be attributed to participation in the FFS? Without a before-and-after assessment of the same farmers, it is difficult to answer with absolute certainty; but the cross-sectional comparison between participants, non-participants and controls offers strong indications of the effects of the FFS.

Insect-control practices

One of the major objectives of the FFS is to give farmers the skills and confidence to reduce their reliance upon insecticides. A comparison of the practices of FFS participants with other farmers indicates that the training is responsible for some important changes (see Table 7.3).

FFS participants make significantly fewer insecticide applications than their neighbours. The difference is clear for the most recent *Maha* season, and for the total applications during the past two and three seasons, respectively. Insecticide practices can vary from year to year and it would have been useful to obtain historical data; but most farmers could not recall such information. Indeed, about one fifth could not recall beyond the most recent *Maha* season.

There is some variation in insecticide practices between locations. Insecticide use is very low in location 1 (the location with the highest yields, where farmers are particularly aware of the most efficient rice management practices) and almost non-existent in location 7 (the tea-growing area where farmers invest little time or interest in rice cultivation). Thus, low insecticide use on rice may represent opposite extremes: careful attention to the latest technology, on the one hand, or lack of attention to the crop, on the other. Despite such differences, FFS participants exhibit markedly lower insecticide use than non-participant neighbours in each location.

There are two principal components of this difference in the number of applications. First, there are more FFS participants than non-participants who apply no insecticide. In the most recent *Maha* season, 60 per cent of FFS participants used no insecticide. However, an appreciable number of non-participants and control farmers made no insecticide application during that season (34 and 40 per cent, respectively). Second, participation in FFS seems to eliminate high numbers of applications. Only 7 per cent of the FFS participants made more than two applications of insecticide during the most recent season, compared with 27 and 29 per cent of non-participants and controls, respectively. None of the FFS participants made more than four applications.

Table 7.3 *Insect-control practices of farmer field school (FFS) farmers and their neighbours*

Practice	FFS farmers (n = 70)	Neighbours (n = 70)	Statistical significance of difference between FFS and neighbours	(Control farmers) (n = 70)
Insecticide applications, Maha 2002/2003	0.6	1.7	<0.001	(1.7)
Total insecticide applications, Maha seasons 2002/2003 and 2001/2002	1.0 (n = 57)	2.8 (n = 57)	<0.001	(3.0) (n = 58)
Total insecticide applications, Maha 2002, Yala 2002, Maha 2001/2002	1.5 (n = 56)	3.8 (n = 57)	<0.001	(4.0) (n = 57)
Report trend of lower insecticide use (percentage farmers)	80	49	<0.001	(49)
Report increased time in monitoring (percentage farmers)	43	12	<0.05	(12)
Hours monitoring fields [a]	5.2	5.3	ns	(5.7)
Apply herbicide, Maha 2002/2003 (percentage farmers)	99	100	ns	(99)
Apply fungicide, Maha 2002/2003 (percentage farmers)	21	39	<0.05	(29)

Note: [a]Monitoring time is calculated as hours per acre per week. The averages do not include a few outliers (about 5 per cent of the total sample) that provided exceptionally high estimates, all of which involve fields of 0.5 acres or less.

It is possible that the FFS participants were using less insecticide before they enrolled in the course; but a higher proportion of these farmers say that their insecticide use has dropped in the past decade than do non-participants or controls. FFS participants were asked to estimate their average insecticide application rates before and after the FFS. Their recall of practices before the course was 3.1 applications per season (which may be an exaggeration).

All rice farmers make frequent visits to their fields throughout the growing season to monitor pest and disease damage, check on water levels and irrigation conditions, and assess the growth of the crop. More of the FFS farmers than their neighbours report that they are spending an increasing amount of time in monitoring. When asked specifically to compare pre- and post-FFS practices, about one quarter say that they have increased their frequency of inspection and one half estimate (mostly small) increases in time per visit. On the other hand, a calculation of hours per acre based on farmer estimates of frequency and duration of field visits does not show any significant difference in monitoring time between FFS farmers and others.

The survey also revealed that FFS farmers use significantly less fungicide than other farmers. The majority of farmers could not name the specific disease for which they applied fungicide; the relatively few FFS farmers using fungicide were slightly more articulate on this subject than the others.

Although FFS farmers use significantly less insecticide and fungicide than other farmers, their weed control practices are essentially indistinguishable from those of other farmers. Virtually all farmers use herbicide; most farmers rely upon a single application and FFS farmers are slightly less likely to be among those who use two or more applications.

Insect-control knowledge

For those FFS farmers who continue to use insecticides, there are few differences from other farmers in terms of knowledge or target pests. A surprisingly high proportion (about one third) of farmers were unable to name the specific insecticide(s) that they used during the previous season, and FFS farmers who used insecticide were no more likely than other farmers to be able to name the specific product. By far the most commonly reported insecticide for all groups is fenobucarb, a carbamate insecticide. The large degree of uncertainty regarding product names makes analysis of insecticide types problematic, but FFS farmers reported a somewhat lower reliance upon organophosphates than their neighbours or controls. The most common target pest for all insecticide use was the brown plant hopper, accounting for about 40 per cent for all groups of farmers.

The survey asked farmers several questions about their knowledge and opinions on insecticide use. Table 7.4 summarizes those instances where there were interesting differences between types of farmers. The FFS devotes some time to building up farmer knowledge of natural enemies of the most important rice pests, and FFS graduates were able to name a significantly greater number of these than could other farmers. The questionnaire also examined basic decision rules for insecticide application (asking farmers whether they always, sometimes or never follow a particular rule). A few differences stand out. FFS farmers are much more likely to say that they never apply insecticide simply on the basis of visual evidence and are more likely to insist that insecticides are used only after some type of quantitative assessment of damage. For those farmers who reported a trend towards lower insecticide use, the FFS farmers were more likely to attribute this to an effort to promote natural pest control and less likely to say that it was simply in response to lower overall pest incidence. A particularly important issue in reducing insecticide dependence is building up farmers' confidence that the rice plant is able to compensate for early leaf damage, which does not warrant insecticide application. FFS farmers showed that they understood this relationship better than their neighbours.

Table 7.4 *Farmers' knowledge and opinions on insect control*

Knowledge or opinion	Farmer field school (FFS) farmers (n = 70)	Neighbours (n = 70)	Statistical significance of difference between FFS and neighbours	(Control farmers) (n = 70)
Number of natural enemies named 3.8		1.5	<0.001	(1.6)
Decision rule for insecticide use (percentage farmers)				
'Never on evidence of attack alone' 81		40	<0.001	(61)
'Always after counting insects or damaged plants'	74	31	<0.001	(51)
If trend is towards less insecticide, why? (percentage farmers)				
'Promote natural pest control'	80	56	<0.05	(35)
'Fewer pests'	4	21	<0.05	(35)
Opinions about insecticide (percentage farmers)				
'Insecticide use increases yield'	13	27	<0.1	(26)
'Leaf-feeding insects in early stages cause severe damage'	33	57	<0.01	(43)
'You must spray early to control leaf-feeding insects'	13	47	<0.001	(39)

The costs of reduced insecticide use

Does the change in insecticide use (which reduces cash outlay for chemicals and often for labour) involve any additional cost? The FFS emphasizes understanding ecological relationships in order to avoid unnecessary insecticide use, rather than teaching alternative control methods. The information in Table 7.4 indicates that FFS farmers' attitudes and perceptions are different from those of their neighbours. But the fact that insecticide use can be reduced without yield loss is due in large part to the re-establishment of regulatory processes in the rice ecosystem that were compromised by high insecticide use. There is no evidence that FFS farmers consistently invest time or money in any substitute techniques for insect control.

The only possible exception is time for field monitoring. Table 7.3 indicated that FFS farmers are more likely than others to say that they are spending more time in monitoring; but their actual estimates of hours in the field show no differences from other farmers. Such estimates can only be approximate and, in any case, reflect time spent in various tasks in addition to monitoring pests. During informal conversations some FFS graduates say that, because their monitoring is now better informed, they spend no more time, or even less time, than previously. Our conclusion is that participation in the FFS apparently gave these farmers enough additional knowledge and confidence to forgo unnecessary applications of insecticide without requiring a significantly higher time investment to counterbalance the savings on purchased inputs.

Soil fertility practices

The FFS devoted a significant amount of time to other subjects as well, and it is important to see if this investment translates to other changes in crop management. The study paid particular attention to soil fertility management. Table 7.5 summarizes the most important survey results for this subject.

The FFS (and a number of other extension activities) encourage farmers to switch from pre-blended fertilizers (known as 'paddy mixture') to single nutrient fertilizers, particularly triple superphosphate (TSP) and muriate of potash (MOP) (along with urea as the source of nitrogen). The state fertilizer corporation has stopped producing such mixtures; but private fertilizer companies provide blends and some input dealers offer farmers their own mixtures. Some farmers still find it more convenient to buy a mixture than to learn about the individual components. More FFS farmers than others have switched in the direction of more precise fertilizer usage. Extension (and the FFS) also recommends the use of zinc sulphate; but very few farmers of any type have yet adopted this practice.

Table 7.5 *Soil fertility practices of farmer field school (FFS) farmers and their neighbours*

Practice (percentage of farmers)	FFS farmers (n = 70)	Neighbours (n = 70)	Statistical significance of difference between FFS and neighbours	(Control farmers) (n = 70)
Use paddy mixture	24	50	<0.01	(50)
Use triple superphosphate (TSP)	83	51	<0.01	(54)
Use muriate of potash (MOP)	83	53	<0.01	(56)
Incorporate rice straw	86	73	<0.1	(60)
Use compost	1	0	ns	(0)
Use animal manure	13	7	ns	(9)
Use green manure	14	10	ns	(11)

Another important extension message of the past decade has been the value of incorporating rice straw after the harvest, rather than burning it. This involves spreading the straw on the field after the first ploughing, then inundating the field and allowing the straw to partially decompose before the second ploughing. Farmers estimate that the practice requires approximately five additional days of labour per hectare.[6] The survey showed that straw incorporation is more widely practised in the study locations than in many other areas of the country (van den Berg et al, 2002) and that FFS farmers are slightly ahead of the others in adopting this practice.

The FFS also encouraged farmers to place more emphasis on other organic fertility sources, such as green manures; but there is no evidence of any change in these practices because of the FFS. The most common green manure source is *Gliricidia sepium*; but only a handful of farmers use it on their paddy fields (in the same way as straw). One of the principal drawbacks is the labour required to carry the cuttings to the field. Farmers point out that it is those who have paddy fields on the edge of the *yaya*, close to land where such crops might be grown, that are most likely to take up this practice.

Farmers were asked about their attitudes towards the use of synthetic fertilizers. The majority agree that it is unwise to depend completely upon these fertilizers, and there is no difference in opinion across groups. The survey also asked farmers to describe their trends in fertilizer use. A higher proportion of FFS farmers than their neighbours (41 versus 19 per cent) say that they are using less fertilizer now than in the past (however, 29 per cent of the FFS farmers, and 54 per cent of their neighbours, say they are using more fertilizer). For those who cite a downward trend, the FFS farmers are most likely to give the additional use of organic matter (principally rice straw) as the reason, while their neighbours' most common explanation is the high price of fertilizer.[7]

Activities subsequent to the FFS

A major reason for studying farmers who had participated in a FFS five or more years ago was to assess the degree to which this experience was responsible for any subsequent individual or group action. Although many FFS graduates ask for similar training to help reduce insecticide use in vegetables (where frequency of insecticide application is generally much higher than in rice), a comparison of the insecticide practices of the relatively small number of respondents who grow vegetables on a commercial scale indicated no consistent differences between FFS farmers and others with respect to insecticide use. A few FFS farmers reported trying local insect-control methods for vegetables, in addition to chemicals.

The FFS encourages farmers to do their own experiments, and expects that participants will increasingly test new ideas and take the lead in innovation. The survey asked FFS farmers to describe any experiments they had done subsequent to the FFS; but only 8 out of 70 could give any examples, mostly concerned with trying an organic fertilizer or changing ploughing depth. All farmers were asked to describe any innovations that they had tested in crop management; but so few farmers of any type were able to give examples that further analysis was not worthwhile. Thus the survey provides little support to the belief that FFS participation leads to increased experimentation and innovation.

The survey also examined possible differences in attitude to farming among the different types of respondent. As mentioned above, FFS farmers were more likely than their neighbours to participate in external agricultural courses and events, perhaps because of relationships formed with extension. All farmers were asked if they wanted at least one of their children to become a farmer. Overall, slightly less than half of the respondents said 'yes', and FFS farmers were not significantly more committed than others to seeing that their children became farmers (it is noteworthy that the larger farmers in the sample are somewhat *less* likely than others to express an interest in their children becoming farmers).

Some descriptions of FFS techniques hold out the hope that a successful experience will lead to other group activities, emphasizing that the FFS can build social capital. There was little evidence of this in the study. Of the seven FFS we examined, the only subsequent group activity reported was one case of participation in rice seed production, a widespread activity organized by extension in many villages in southern Sri Lanka.[8]

One of the most commonly mentioned examples of FFS-inspired activity in Sri Lanka is the development of pesticide-free rice marketing (see, for example, van den Berg et al, 2002). However, the idea is much more widely discussed than actually practised. The major example where this activity has taken hold is Monaragala District in Uva Province. Two local merchants, with assistance and encouragement from a Norwegian-funded integrated rural development project, have begun to purchase pesticide-free rice from local farmers for sale in distinctive packaging in Colombo supermarkets. The several *yayas* contracted by these merchants had been sites of previous FFS. The project shows initial signs of success; but it depends upon the investments of a donor project and local grain traders rather than upon the independent initiative of FFS farmers. As word spreads of the premium paid for pesticide-free rice by these traders, more farmers are interested in joining; but it will take several years to assess the limits of the market. It would appear that the most difficult challenge for farmers joining this scheme is to eliminate herbicide use rather than to give up insecticides.

Another frequently mentioned product of FFS is IPM clubs. Again, we were not able to find any evidence that these represent farmer initiative. The FAO project provided encouragement, and a small amount of support, to establish such clubs. The members we interviewed in two locations indicated that the clubs meet infrequently and are dependent upon the DoA. Club members participate in occasional visits to DoA-funded events or 'IPM congresses'. When asked what an IPM club is for, members say that they hope to gain access to credit for inputs, machinery or better marketing conditions for rice; but the conversations we held revealed no independent initiatives or direct links to FFS activities. As one farmer remarked: 'We don't talk about IPM in the IPM club.'

The diffusion of knowledge

We have established that the FFS were responsible for some important changes in the practices and knowledge of participating farmers. To what extent did these ideas spread to other farmers? This depends upon the nature and extent of farmer-to-farmer information exchange. One of the principal criticisms of the FFS strategy has been its relatively high expense per participant, and one of the most common defences is that participating farmers are likely to pass some of the knowledge gained to others. The study looked at the nature of information transmission linked to the FFS.

Comparison of non-participants and controls

One way of examining the extent of information diffusion is to compare the behaviours and attitudes of the non-participant neighbours with those of control farmers. The argument here is that information would be more likely to pass to neighbour farmers in the same *yaya* than to those several kilometres away through conversations and direct observation. Thus, if information diffusion took place we would expect the neighbours' behaviour to be closer to that of the FFS participants than that of the control farmers. Some of these comparisons can be made by referring to the data in previous tables. Table 7.3 shows that there is essentially no difference in insecticide practices between neighbours and controls (except that neighbours apply slightly more fungicide than do the controls). Table 7.4 indicates that neighbour farmers' decision rules and rationale for insecticide reduction are intermediate between those of the FFS participants and controls, but their knowledge of natural enemies and their opinions about early spraying are no more enlightened than those of the controls. Table 7.5 shows that although a slightly higher proportion of neighbours than controls incorporate rice straw, other soil fertility practices are indistinguishable between the two groups. Thus, on the basis of these comparisons there is only slight evidence that what FFS farmers learned was transmitted to other farmers.

Sources of information

A second way of approaching the issue of information flow is to ask farmers how and when they heard about particular practices. Like the rest of us, farmers have imperfect memories and so there are limits to the specificity of recall that can be expected. But the questionnaire allowed at least an initial exploration of how farmers acquired information about lower insecticide use, straw incorporation and the use of green manure.

There have been various efforts by extension urging farmers to reduce insecticide use. Relatively more of the non-participants (56 per cent) than the controls (36 per cent) are aware of these campaigns; but the difference between the two groups is entirely due to the greater knowledge of the FFS among the former (who are more likely to have seen their neighbours attending a FFS). Table 7.6 compares those farmers from all the three groups who say that they have reduced their insecticide use. The participant farmers, not surprisingly, give most credit to the FFS. Other farmers tend to cite either formal extension activities or their own ingenuity, and very few say that other farmers were the source of information on this topic. Whether this analysis accurately reflects the actual level of farmer-to-farmer transmission of information is unclear; but at a minimum we can say that (in response to a questionnaire) farmers are not likely to credit other farmers with providing the inspiration for changing their insecticide practices. This may have implications for what can be expected from farmer-to-farmer transmission of FFS messages.[9]

Table 7.6 *Farmers' sources of ideas for lowering insecticide use (percentage of farmers)*

Type of farmer	Own idea	Another farmer	Farmer field school (FFS)	Demonstration or Department of Agriculture advice	Other
FFS (n = 56)	9	0	86	4	2
Neighbour (n = 33)	42	15	3	33	6
Control (n = 34)	35	0	0	44	21

Those farmers who incorporate straw were asked when they began using this practice. The responses indicate a fairly gradual process of adoption covering more than a decade, although 29 per cent of FFS farmers and 54 per cent of the others adopted in 2000 or later. It is not possible to demonstrate significant adoption by FFS farmers immediately after the FFS course; indeed, 38 per cent of the FFS farmers were incorporating straw before they participated in the course.

A total of 31 farmers listed 38 instances of the use of green manures (currently or during the past), another technique promoted by the FFS (and by other extension activities). FFS farmers are no more likely than others to use green manures, so it is not useful to consider them as important information sources for this practice. In only two cases did farmers say that they had learned the practice from another farmer; the most common response (61 per cent) was a DoA demonstration.

Learning from the FFS

A third way of examining information diffusion is to try to follow the learning that took place in the FFS and subsequent transmission to other farmers. When asked what new things they learned from the FFS, most participants have little trouble in listing four or five items. The questionnaire asked the FFS farmers to list the major items in order of priority. If we divide the responses into broad categories (insect control, soil fertility, etc.), examples of insect control are the most common first response, followed closely by examples of soil fertility. If we examine the total of responses to this question (limited to the top five responses), soil fertility practices are slightly more frequent than insect control (40 versus 36 per cent), followed by planting practices (24 per cent, including such practices as the importance of deep ploughing and the selection of adequate seeding rates). Whether soil fertility gains first place because it is seen as more important by farmers, or simply because there are more soil fertility techniques to learn, is open to question. When we asked FFS farmers what they are now doing differently because of the course, answers related to soil fertility were more frequent than insect control techniques (46 versus 33 per cent). Again, this is partly due to the fact that there are several soil fertility innovations available for adoption, while pest control innovations are largely limited to lower insecticide use and more careful monitoring.

The questionnaire attempted to obtain specific details about the number and type of farmers to whom FFS graduates passed information, but it was only partially successful. Not surprisingly, there was a wide range of responses with respect to types of information transmitted and numbers of recipients. On average, FFS farmers claim that they told about 10 relatives and 19 other farmers about their experiences; but there is wide variation and no clear pattern regarding what type of information or recipient is most important.

FFS farmer claims of information transmission can be compared with reports from other farmers. Most (74 per cent) of the neighbour farmers had heard of the FFS; the only exception is location 7, the tea-growing area, where there was very little awareness among non-participant farmers. A minority (26 per cent) of control farmers had also heard of the FFS. Most of the farmers who had heard of the FFS said that they learned things from farmers who had participated, and we asked them to list specific instances. Table 7.7 summarizes some of the responses of neighbour farmers (there were too few responses from controls for meaningful analysis). Examples of soil fertility practices were more important than those related to insect control. Straw incorporation is by far the most prevalent, and most farmers who learned about it say they followed through with its adoption. The next most frequent response was the use of single-nutrient fertilizers, although a smaller proportion of farmers say they changed their practices accordingly. Although a number of the farmers receiving information described various types of pest-control knowledge gained from conversations with FFS farmers, only a relatively few (11 per cent) mentioned actual reduction in insecticide use as a consequence of this knowledge. Only 6 per cent of neighbour farmers say that they heard something about the importance of beneficial insects from FFS farmers.

Table 7.7 *Most frequent instances of information transmission from farmer field school (FFS) farmers to their neighbours (n = 70)*

Information	Number (and percentage) of neighbours who received the information from FFS farmers	Number (and percentage) of neighbours who acted on the information
Insect control		
Use of insecticides only as last resort	8 (11%)	6 (9%)
Importance of beneficial insects	4 (6%)	4 (6%)
Soil fertility		
Use of straw as soil amendment	21 (30%)	19 (27%)
Use of single-nutrient fertilizers	10 (14%)	5 (7%)
Land preparation		
Land levelling and deep ploughing	7 (10%)	6 (9%)

The analysis is complicated by the fact that most of these practices have been promoted through other extension strategies in addition to the FFS. Thus, farmers may confuse their sources of information. But many farmers seem familiar with the concept of the 'field school' (*shashthra pasala*) and are able to distinguish it from other extension activities. In a study describing an evaluation of a FFS programme run in Sri Lanka by CARE, K. Jones (2002) reports that many non-participants acknowledged that they had learned things from FFS farmers. Our study confirms this observation in so far as about half of the farmers in the same *yaya* are aware of some of the activities of the FFS, but cautions that what these other farmers report is not necessarily related to pest control, nor is it always translated into actual practice.

To summarize, FFS participants are generally enthusiastic about their experience and many are apparently eager to communicate their knowledge. About half of the neighbouring farmers could report at least one piece of information received from the FFS graduates. However, not all of the information received was acted upon, and by far the most common type of information discussed was related to soil fertility. Although the totals are relatively modest, the number of farmers who claim to have taken up straw incorporation because of what they learned from FFS farmers is three times the number who credit the FFS farmers as a source of information for lowering insecticide use. It is relevant to consider that straw incorporation is a practice that a farmer can observe his neighbours doing. The rationale for reducing insecticide use, on the other hand, is more difficult for a neighbour to appreciate.

Factors related to the use of new production practices

The emphasis in this chapter has been on assessing the degree to which farmers who attended a farmer field school learned new concepts and changed their practices, and the degree to which this experience influenced participating farmers to initiate other activities or contributed to interactions with other farmers. It is also possible to examine the same data from the perspective of technology adoption, without reference to the FFS. This section briefly considers what aspects of information access and labour patterns might be related to the adoption of low external-input technology (LEIT) for pest control and soil fertility management (lowered insecticide use, straw incorporation and the use of green manures).

Information

There is no evidence that the adoption of any of these practices is associated with higher levels of education. Education levels in the sample are quite high, and the new practices do not require any particularly specialized knowledge.

Although formal education is not an important factor in these cases, certain types of knowledge may influence production practices. We have seen that FFS farmers have certain opinions and decision rules that distinguish them from other farmers. In addition, when asked to give their reasons for reduced insecticide or fertilizer use, FFS farmers are more likely than others to cite environmental (rather than economic) reasons.

Table 7.8 explores the relationship of certain beliefs and knowledge to insecticide use. It shows a significant correlation between what non-FFS farmers believe about the damage caused by leaf-feeding insects and consequent insecticide use. Those who see leaf-feeding insects as a particular threat spray significantly more than others. On the other hand, there is no correlation between ecological knowledge (number of natural enemies named) and number of insecticide applications for non-FFS farmers. Such knowledge is certainly valuable, but may by itself have little influence on farmer behaviour.

Table 7.8 *Relationship of beliefs and knowledge to insecticide use*[a]

Belief or knowledge	Number of insecticide applications during past two Maha seasons	Statistical significance of difference
Belief about early spraying		
Need to spray early	4.2 (n = 49)	<0.001
Don't need to spray early	2.0 (n = 63)	
Knowledge of natural enemies		
Can identify two or more	3.2 (n = 57)	ns
Can identify one or none	2.6 (n = 58)	

Note: [a]Non-participants and controls only.

The study did not examine farmers' environmental knowledge or beliefs in any depth; but there would appear to be a growing consciousness of the problems caused by insecticides and by over-reliance on synthetic fertilizers. However, there is little evidence for the emergence of any type of 'green' farmer. Figure 7.2 summarizes relationships between various soil fertility practices and insecticide use. Those farmers who are moving toward more precise soil fertility management by purchasing single-nutrient fertilizers use considerably less insecticide, perhaps in part representing a growing sophistication about purchased inputs in general. Perhaps surprisingly, those farmers who incorporate straw (a 'green' practice) use more insecticide than other farmers.[10] There is no relationship between green manure use and insecticide applications.

We could find no relationship between membership in formal organizations and the adoption of the new insect-control and soil fertility management practices. The practices described in this study do not require any cooperation or coordination among farmers. Informal links among farmers certainly contribute to the spread of information, and the previous section indicated that certain innovations from the FFS were communicated to other farmers. Nevertheless, there is relatively little evidence of farmer-to-farmer spread of some of these techniques. It is important to recall that farmers are unlikely to say that they learned about lower insecticide use or green manures from other farmers. On the other hand, when we asked about alternative pest-control

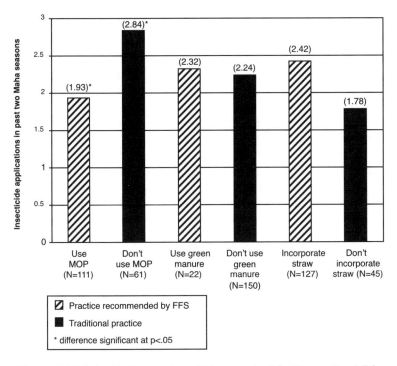

Figure 7.2 Relationship between insecticide use and soil fertility practices (all farmers)

practices (such as water control in response to brown plant hopper or the use of wood ash for leaf folders), respondents were relatively more likely to cite other farmers as the original source of knowledge; but even in these cases, 'another farmer' never accounted for as much as half the responses.

There is somewhat more evidence for farmer-to-farmer transmission of knowledge about straw incorporation. The practice is particularly visible, and it appears to make an important contribution to yield. In our overall sample, farmers who incorporated rice straw had yields 38 per cent higher than the others. The yield response to straw incorporation in irrigated rice varies according to several factors; but a comprehensive review estimates an average yield gain of about 0.4 metric tonnes per hectare (Ponnamperuma, 1984). Our crop management data are inadequate for a thorough analysis (particularly in the absence of rates and timing of fertilizer applications). Nevertheless, several regressions were run with and without locations 1 and 7 (the yield extremes). We regressed the yield data on management practices (for example, pest and disease control, and types of inorganic and organic fertilizer) and farmer assets (for instance, education, FFS participation and farm size). The only consistently significant factor in all the regressions was straw incorporation (FFS participation, on the other hand, was *never* a significant determinant of yield in these regressions).

Labour

Some LEIT may require a significant investment of time, which affects the probability of adoption. But the reduction of insecticide use is a cost-saving practice and it does not appear to be accompanied by a significant increase in monitoring time (although the skill and knowledge with which fields are monitored may increase). Straw incorporation requires a modest increase in labour; but this comes at a time when the farmer is likely to be present supervising land preparation, and the apparent yield advantage makes it a widely adopted innovation. Green manures are not widely enough promoted or utilized to allow analysis; but they certainly require more labour (preparing cuttings, carrying them to the field and preparing them for incorporation). In some cases, farmers take the cuttings from plants growing on common land; but in other cases they establish their own plantings of a green manure crop, which requires additional time. The potential yield contribution may be important; but only a few farmers have yet decided that the investment is worthwhile.

Household livelihood diversity

Although the households in the sample depend upon various income sources, this livelihood diversity does not appear to be a particularly important determinant of technology adoption. Three explanations can be put forward. First, nearly three-quarters of our sample describe themselves primarily as farmers, and most of the other occupations offer sufficient flexibility to maintain some control over the basic operations of the paddy field. Our sampling served to eliminate areas where rice cultivation takes second place to off-farm employment (partly because we were concerned to study well-attended FFS). Second, although the farms are small, most employ at least some hired labour. Third, as we have seen, most of the technologies examined do not require a significant time investment.

Nevertheless, to the extent that the new practices require additional time and attention, the deployment of household labour resources may have a bearing on acceptability. A few relationships between household labour strategies and technology use are explored in Table 7.9. The sample sizes of the occupational groups are not large enough to warrant statistical tests, but a few trends emerge. Those who say that their principal occupation is casual labour are likely to be working, or searching for work, on a fairly continuous basis and have less time to devote to farm management. Despite their small holdings, they apply insecticides more frequently than others and they are less likely to have learned about single-nutrient fertilizers or to use organic amendments. Most of the farmers who also work as agricultural labour do so only occasionally, and their practices (and income patterns) are closer to those of other farmers, except that they use a higher amount of insecticide (we noted earlier that these two labouring groups are less likely to take an interest in the FFS). Those with salaried jobs or who are engaged in trade generally maintain an interest in their paddy fields and are as well informed as full-time farmers about new practices, including their participation in FFS.

Thus, household livelihood diversity seems to cut in two directions with respect to the adoption of new practices. On the one hand, farmers who must also do casual or farm labour have less time for activities such as a FFS or to learn about new practices. On the other hand, farmers whose skills and resources open up salaried or business opportunities still have enough time and interest in their rice fields to stay abreast of the latest innovations.

Table 7.9 *Principal occupations and cultivation practices*[a]

Factor	Farmer[b] (n = 103)	Farmers who work as farm labour (n = 30)	Business or skilled (n = 17)	Salaried (n = 17)	Casual labour[c] (n = 12)
Participate in farmer field school (FFS)	37%	20%	47%	41%	8%
Number of insecticide applications during past two Maha seasons	2.6	3.6	1.8	2.1	3.6
Use muriate of potash (MOP)	70%	67%	82%	71%	25%
Incorporate straw	75%	87%	77%	71%	58%
Use green manure	14%	7%	0	12%	8%
Total paddy area (hectares)	1.1	0.8	1.0	0.9	0.5
Cash income from rice	53%	51%	30%	32%	36%
Years of education	6.9	6.1	8.5	9.1	7.2

Notes: [a] All farmers except location 7.
[b] Excluding those who work as farm labour.
[c] Not an occupation category (an activity performed in addition to principal occupation).

Summary

The FFS programme in Sri Lanka has had a significant impact on the farmers who participated. FFS farmers use only one third as much insecticide as other farmers and they are less dependent upon fungicides. This difference in practices is evident five or more years after the conduct of the FFS. FFS farmers also demonstrate a better appreciation of the agro-ecological relationships that affect pest control. The FFS not only introduced farmers to more rational insecticide use, but also promoted various other crop management techniques. FFS farmers are farther ahead than their neighbours in the move towards straw incorporation for soil fertility enhancement and the use of single-nutrient fertilizers. Although it is well to remember that the study deliberately selected FFS sites that had been well managed, it is likely that these results would be evident for the majority of FFS sites in the DoA programme.

The practices promoted in the FFS that have had the widest acceptance are of immediate relevance to farmers for yield enhancement and/or cost saving. They are applied to the management of a crop that provides at least self-sufficiency for almost all the farmers and is an important source of cash income for the majority. Given this context, it is not difficult to attract farmers' attention (the relative lack of interest in the tea-growing area is the exception in the sample that proves the rule). Most farmers are able to attend a FFS if the opportunity arises. Although FFS participants tend to have slightly above-average resources, this is not an important determinant of participation or learning. The major problem appears to be for farmers whose off-farm wage labour interferes with their ability to attend a season-long course.

The situation of the rice growers we examined in southern Sri Lanka is probably more socio-economically homogeneous than some other environments in Asia where FFS have been conducted. The range of landholding in the sample is not great, and those farmers who are tenants have secure access to their plots. A correlate of this relative uniformity in rice-farming assets is the minority of instances of non-farm activities that influence farmers' interests and capacities regarding opportunities such as the FFS. Thus, the relative equality of participation in the FFS may not necessarily hold for other countries or other areas of Sri Lanka.

Although the FFS was meant to introduce IPM, it is difficult to talk about farmers 'practising IPM'. This study has emphasized specific practices and knowledge, rather than the more general concept of IPM. On the one hand, there is evidence that FFS participants learned enough about certain ecological relationships to give them the confidence to reduce their dependence on insecticide; farmers come away from the FFS with more than a few simple rules. But on the other hand, it is not clear that this knowledge is applied to anything other than a more informed approach to insecticide use. We have no evidence that farmers are consistently applying additional or novel practices for pest control.

Despite the significant effects of the FFS on the participants, it is difficult to demonstrate much outward information flow. Many of the participants show considerable enthusiasm about what they learned and an eagerness to communicate; but relatively little of this shows up in the practices of other farmers. The analysis is complicated by the fact that most of these techniques have been promoted through various extension strategies beyond the FFS, so it is sometimes difficult to identify the original source of information. But there is little evidence of significant farmer-to-farmer transmission. In most cases, farmers are more likely to cite extension or other formal activities as the source of their information, rather than crediting other farmers. Those practices that are more visible (such as straw incorporation) seem to be more likely to be taken up by others. The message of lower insecticide use is apparently more difficult to transmit, and this is complicated by the fact that farmers' confidence in reducing insecticide usage rests, in part, on an understanding of ecological relationships that is difficult to communicate. This is not to say that farmer-to-farmer transmission is unimportant; but this study was unable to demonstrate that particular examples of LEIT are diffused to any significant degree by local social networks.

There is a fairly widespread belief that participation in a FFS can lead to important advances in participants' capacities (beyond particular crop management techniques) and can contribute to further farmer organization. The study found little evidence for this. Although FFS farmers appreciate the experimentation offered in the course, they do not usually go on to perform further experiments or engage in innovatory behaviour that sets them apart from their neighbours. Similarly, the study found no significant examples of follow-up activity, and reported cases of such activity appear to depend upon the resources and priorities of external projects rather than upon the initiative of FFS farmers.

Several of the examples of LEIT featured in the FFS are achieving widespread adoption. Basic parameters of human and social capital (for example, education and membership in local groups) are not determinants of farmers' ability to take advantage of these techniques. Although many farmers appreciate the resource-conserving rationale of these techniques, there is no evidence of an emerging set of farmers who are more likely to practise such techniques because of their environmental advantages or their relation to input self-sufficiency. Decisions on the utilization of these technologies are taken on a case-by-case basis.

The reduction of insecticide use does not appear to require much additional time for field monitoring, and no other pest-control technique has been introduced that might require significant labour inputs. Those farmers who continue to use high amounts of insecticide may not have had the time or the incentives to learn about the possibility of reduced dependence; but if they do so, their additional labour investment will be minimal. Straw incorporation requires a modest amount of additional labour; but it would appear that the yield advantages repay the investment. Only those farmers who spend a significant amount of time in off-farm work appear to be slower in taking up these techniques. In contrast, green manure utilization requires more labour and learning, and is not yet widely practised.

With respect to the promotion of LEIT, the Sri Lanka FFS case offers both hope and challenges. On the one hand, the FFS programme is an obvious success and has contributed to the growing utilization of several resource-conserving practices. The experience shows that it is possible to lower the use of insecticides and to take better advantage of organic soil fertility resources. On the other hand, the next steps are less certain. Reducing the excessive dependence upon insecticides is possible by re-establishing the natural resilience of the irrigated rice ecosystem. But pest damage is still a problem and it will be necessary to identify additional control measures, most of which will require some labour or cash investment. Reintroducing farmers to the practice of incorporating straw rather than burning it is made easier by the removal of the fertilizer subsidies that occasioned the abandonment of the practice in the first place. But most other soil amendments that can be offered require considerable labour investment. And finally, it is necessary to explore how new resource-conserving crop management techniques can be developed and diffused most efficiently where there is little evidence of autonomous farmer-to-farmer information transmission.

Notes

1   The field research was carried out with advice and guidance from the staff of the Department of Agriculture in Matara and Hambantota districts. We acknowledge the dedication and thoroughness of the survey enumerators (I. K. Ruwansiri, J. C. Ratnayake and D. N. D. K. Mendis). Earlier drafts of this chapter benefited from comments by Janny Vos, Henk van den Berg, and Michael Richards.

2   See K. Jones (2002) for a description of a large FFS programme in Sri Lanka organized by CARE.

3   This type of tenancy is very common in Sri Lanka. Before the Paddy Lands Act of 1958, tenants gave as much as three-quarters of the harvest to the landowner; but the act provided security of tenure and established maximum tenancy fees.

4   Moore (1985) suggests that one of the reasons that Sri Lankan smallholders do not form a unified political grouping is their high dependence upon non-farm sources of income. Spencer (1990) discusses the fact that the importance of *chena* (shifting) cultivation of a variety of crops is often underemphasized because of the importance attached to paddy cultivation, even though many farmers have little or no paddy land.

5   The use of hired labour in paddy production is not a recent phenomenon. A study in Hambantota District during 1972 to 1973 recorded that 67 per cent of households relied at least partially upon hired labour for sowing, 44 per cent for weeding, 37 per cent for pesticide application and 79 per cent for harvesting (Chinnappa and Silva, 1977).

6   See Ali (1999) for a discussion of the trade-offs between organic fertilizer use and labour costs in Asian rice.

7   Most government subsidies on fertilizer have been removed during the past few years.

8   The government has withdrawn from large-scale paddy seed production and sale, leasing some of its seed farms to private firms in the hopes of stimulating private seed supply. At the same time, many DoA offices organize local-level seed production activities.

9   A similar result was noted in the Philippines by Rola et al (2002).

10   There is no immediate explanation for this relationship, which may be coincidental. Rice straw incorporation may be implicated in the spread of certain diseases; but this would not explain higher than average insecticide use.

8

The Trajectory of Low External-Input Agriculture

Robert Tripp

Introduction

The first chapter of this book presented a set of questions about the performance of low external-input technology (LEIT) that the field research of the three case studies (complemented by a review of the literature) set out to address. This concluding chapter summarizes what we have learned about LEIT and assesses the consequences.

The first part of the chapter reviews case study experience on a series of issues related to the performance of LEIT. These include the extent of farmer utilization of LEIT introduced through project activity; the types of farmers who participate in LEIT projects; the types of farmers who are able to incorporate LEIT in their farming practices; the degree to which LEIT is labour-intensive and the implications for uptake; the degree to which LEIT is knowledge-intensive and the implications for uptake; the aspiration that LEIT utilization can bring about a change in the mindset of farmers and lead to fundamental shifts in farming strategies; the relationship between LEIT utilization and an adaptive, experimental approach to farming; the extent to which LEIT is susceptible to farmer-to-farmer transmission; and the relationship between LEIT and social capital.

The second part of the chapter attempts to draw some conclusions from the analysis of the performance of LEIT. It begins by examining the relationship between technology utilization and agricultural development and asks if LEIT actually has characteristics that accord it a special place in agricultural development strategies. This is followed by a review of the challenges in evaluating the performance and role of LEIT. The limited impact of development projects, such as those that promote LEIT, is usually addressed through calls for more astute project design and then for scaling up; the chapter examines the advisability of such strategies. The chapter concludes with some proposals for allowing the promotion and utilization of LEIT to play a more effective role in supporting equitable agricultural development.

The utilization of low external-input technology (LEIT)

The three case studies were chosen partly on the basis of prior evidence of success, so it is not surprising that the fieldwork confirmed that many farmers have taken advantage of various types of LEIT. Nevertheless, it is worth emphasizing that a significant number of farmers found various examples of LEIT to be useful and have made them part of their crop management strategies. It is also worth noting that the farmers continue to use, maintain or refine these technologies five or more years after the projects responsible for introducing them have terminated. On the other hand, given that these are 'best case' examples, it is important to provide a realistic review of the extent and nature of uptake.

In the two larger project areas in Honduras, the use of soil regenerative LEIT on subsistence crops among project participants is fairly modest. Only 20 per cent are currently using in-row tillage (IRT), 24 per cent are using cover crops and 32 per cent maintain some type of live barrier. However, 47 per cent of the participants who grow vegetables are using in-row tillage on at least part of their plot. In addition, it must be remembered that only about half of the farmers counted as project participants in the study took part in a full complement of project activities. The majority of participants at least experimented with some of these techniques, but many later abandoned them.

In the Kenya study, more than half of the participants in the soil and water conservation (SWC) Catchment Approach actually implemented at least one conservation structure on their farm as a result of the project. There were a number of different options – but in the high-potential sites the most common innovation was grass strips; in the low-potential sites unploughed strips, or strips of aloe or sisal, were most common. Few farmers constructed any major structures, such as the various types of terrace that were demonstrated.

In Sri Lanka, most of the participants of farmer field schools (FFS) significantly lowered their use of insecticide on rice. Almost none of the FFS graduates now use high levels of insecticide and more than half use no insecticide on rice in a given season. The study also indicated that FFS farmers were further ahead than their neighbours in taking up more efficient soil fertility management, including the incorporation of straw and the use of single-nutrient fertilizers.

In addition to examining the proportion of project participants that have taken advantage of various types of LEIT, we must also consider the intensity of usage. For instance, soil conservation techniques may be adopted on all or part of a field. In the Honduras case, a certain minimum area was defined for classifying farmers as adopters of in-row tillage or cover crops, and not all farmers who used these techniques applied them to the entire plot. In Kenya, the principal SWC technique was a variety of vegetative or unploughed strips, and while many farmers established these strips, they tended to be spaced much further apart on the slope than was recommended for effective soil conservation. In the case of insecticide use in Sri Lanka, although FFS farmers used, on average, much less insecticide than non-participants, many of the latter were also conservative in their use of the chemicals. Approximately 30 per cent of non-participants used no insecticide in a given season.

It would be nice to be able to estimate yield or income changes attributable to the uptake of these technologies, but the case study surveys did not allow this type of analysis. Very careful yield data would have to be collected, preferably over several years, including pre- and post-project phases. In the absence of before and after data, an exceptional range of information on farm and farmer characteristics would have to be analysed if we were to attribute yield differences to a particular practice. Otherwise, it would be impossible to estimate the degree to which any yield differences were the result of the new technology itself or, on the other hand, could be explained by the fact that better farmers (or those with better resources) tend to take up new technology. The Honduras study collected data showing that project participants had, on average, maize yields that were one third higher than those of non-participants. The study speculated on the mechanisms through which LEIT could contribute to these higher yields, but could reach no definitive conclusions. Cropping patterns and conditions were so variable in the Kenya case that it was not feasible to analyse yield data. In Sri Lanka, FFS farmers had (non-significantly) higher rice yields than their neighbours. Whether their participation in the FFS had anything to do with the difference is not known; but it is possible to say that the farmers who lowered their insecticide use saved cash, were less at risk from dangerous chemicals and suffered no yield loss.

An appreciation of the contribution of the LEIT projects should also include a review of the situation in which the projects were established. In central Honduras, the use of in-row tillage and cover crops was essentially unknown to local farmers before the first project began, so all of the changes in that area can be attributed to the stimulus provided by the project. The background of SWC in Kenya was more complex, as there had been a number of efforts over several decades to introduce farmers to the importance of soil conservation practices. Nearly half of the SWC structures noted in the fields of sample farmers had been established before the catchment project began. Nevertheless, the project provided important incentives for further activity in this area. In Sri Lanka, there had been a number of previous efforts to encourage farmers to use less insecticide (counterbalanced by pressures from the commercial sector for greater pesticide use). In addition, rice farmers demonstrated a wide range of practices and attitudes with respect to insecticides; a significant minority of non-participants conscientiously avoided insecticides (based on their own common sense assessments). The FFS helped more farmers to appreciate the implications of the overuse of insecticide.

It is fair to conclude from these best-case examples that when LEIT is introduced in a competent fashion under appropriate conditions, farmers are more than willing to learn about, adapt and utilize it. On the other hand, even in the best of cases, the uptake of LEIT is patchy and not as widespread or revolutionary as some advocates would have us believe. LEIT is useful for certain farmers, under certain circumstances, and we should review the major issues related to the utilization of LEIT.

Participation in LEIT projects

When LEIT is introduced through projects, the design of the project helps to determine the opportunities for different classes of farmer to learn about the new techniques. There are many examples in the literature where new agricultural technology is developed or presented in ways that favour better-resourced farmers. Most people who work on LEIT are very aware of these biases and strive to counteract them. For instance, most LEIT is, at least in theory, specifically targeted towards the conditions and interests of resource-poor farmers. A more relevant danger is the tendency of small pilot projects (of any kind) to concentrate on favoured areas or to work with particularly well-motivated farmers whose conditions may not be representative of the poorer majority. LEIT projects are not immune to this type of bias.

The three projects examined here invested significant effort in achieving broad coverage. Each of them covered large areas. The projects in Honduras each worked in 30 to 35 contiguous communities. The projects in Kenya and Sri Lanka were national in scope, although there was considerable discretion left to local extension staff regarding the choice of target communities. In these latter cases, communities had to give evidence that they were interested in the project activity and demonstrate that they could organize in order to participate. In all cases, all farmers within project communities were free to participate.

In Honduras, the non-governmental organization (NGO) projects worked in an open fashion in a number of neighbouring villages. The goal was to cover all communities in an area characterized by considerable rural poverty, and this was achieved with the exception of a few outlying hamlets. There was a wide range of types of participation; some farmers attended only a few meetings or training sessions, while others managed on-farm trials or volunteered as local extension agents. There were some differences noted for those who did not participate; they tended to be older, to belong to fewer social groupings and had less investment in commercial farming and less access to supplementary irrigation. On the other hand, those farmers who emerged as leaders within the project had more education, better access to irrigation and more memberships in local organizations than other participants, but did not have larger farms.

In Kenya, the goal was to provide farm plans for every member of the target communities. The majority of farmers in the catchment villages participated in developing these farm plans for conservation interventions, although not all farmers followed through on them. Once the project was in motion, a number of catchment-level meetings were organized, and attendance at these meetings was quite variable. Those farmers who did not participate are not significantly different from their neighbours, although they depend slightly more upon off-farm income than the participants. The catchment committee members were more active in local groups than other farmers, but did not have farm resources superior to those of other participants.

In Sri Lanka, the FFS was available to only a relatively few villages each season, and there is some indication that extension staff chose villages with which they had worked previously and that had adequate irrigation facilities.

Within a selected village, the FFS was open to any farmer who was interested, although course limitations meant that only a fraction of farmers could participate. Those who did so had slightly larger holdings than their neighbours, but did not have more education or group memberships. However, those farmers who also worked as labourers were much less likely to participate, presumably because of time constraints.

It is difficult to avoid all biases in projects; but many LEIT projects appear to make considerable efforts to reach their intended targets. When projects are available to only a small proportion of villages in a given area, there is a good chance that prior experience or better resources may bias selection towards certain villages; but this is a danger for any project and has no specific relation to LEIT. More relevant examples of possible bias in LEIT projects are a product of the self-selection process at the local level. In methods such as FFS, where a farmer must commit a significant portion of time during the farming season to gain access to the new techniques, those who have less time available (for example, because of other labour demands) may be left out. In many other cases (such as the Honduras and Kenya examples), farmers have a wider range of options for levels of participation, although a certain minimum time commitment may be necessary to gain access to sufficient advice and support to undertake the innovation. The greater flexibility also allows farmers to simply learn more about the innovations and ideas on offer and to begin to form an impression of whether they might be compatible with their own circumstances. Differences in compatibility will be reflected in the actual uptake of LEIT, which we now examine.

The type of farmer who utilizes LEIT

One of the principal sources of support for LEIT is the contention that these innovations are much more likely to be accessible to resource-poor farmers. The argument is that while those technologies that rely upon purchased inputs may be effective, they are often out of the reach of farmers who are engaged in subsistence rather than commercial production, and who do not have access to cash or markets. The implication is that more investment in the development and diffusion of LEIT will help to shift the targeting of agricultural technology generation towards the needs of the poor and help to address the current imbalance in which new technology tends to be utilized by commercially oriented and better-resourced farmers. The review of the literature in Chapter 4 seriously challenged this assumption; there were virtually no cases in which a significant uptake of LEIT was weighted towards the poor and many examples in which the better-off were more likely to take advantage of LEIT.

Despite fairly broad initial participation in the case study projects, there is considerable evidence of differentiation in the actual utilization of LEIT. There was notable unevenness in the extent to which the projects' LEIT techniques were taken up by farmers. The determining factors included the farming environment and farmer resources, livelihood strategies and commercial orientation.

The Kenya case was the only one of the three examples which examined a project working in two contrasting environments. Interest in, and uptake of, SWC practices was much greater in the high-potential area, where crops made a greater contribution to livelihoods. The specific technologies available in the two zones were necessarily somewhat different, and this may, of course, have influenced uptake as well. In Honduras, the study extended to a small area (Guacamayas), which had more potential for commercial agriculture than the two major areas of study and better access to irrigation, and uptake of the project technologies was much greater there.

Farm size was of variable importance in explaining uptake of LEIT in the case studies. In Honduras, farm size itself was not an important determinant of adoption; but those farmers with greater areas of cash crops were more likely to utilize LEIT. In Sri Lanka, the range in farm size is quite narrow and there is no correlation between rice area and insect-control practices. In both the low- and high-potential zones in Kenya, on the other hand, larger farm size seems to be consistently related to the use of SWC (regardless of major sources of household income).

The importance of agriculture as a source of income can also be a determinant of interest in any new farming technology. In Honduras, most farmers were engaged in some off-farm labour; but those farmers in the lowest-resource wealth category had the highest dependence upon off-farm income and were least likely to use either LEIT (or mineral fertilizer). In the high-potential zone in Kenya, there was a fairly clear positive correlation between agricultural income and utilization of SWC, and a negative correlation between income from business or petty trade and the uptake of these technologies. In Sri Lanka, in contrast, farmers who earned a higher proportion of their incomes from salaries or trade were as likely to be interested in integrated pest management (IPM) as those whose income was predominantly from agriculture. Possible explanations for the difference between the Kenya and Sri Lanka cases include the feasibility of balancing the various income-generating activities (for example, petty trade in the Kenya case may keep the farmer away from the field) and perceptions of the profitability of investing in agriculture if other options are available.

It is particularly important to note that LEIT is more likely to be used for commercial farming than for the cultivation of subsistence crops. In Honduras, the project technologies were only used by a minority of farmers on basic grain crops (maize and beans), but in-row tillage was used by almost half the farmers who grew commercial vegetables. Indeed, it would seem that LEIT adoption (initially for grain production) was one of the factors that helped to move some farmers towards commercial vegetable production. In the higher rainfall zone in the Kenya study, those who established SWC structures on their farms earned a higher proportion of their incomes from crop sales. In Sri Lanka, there was no statistical relationship between commercial rice sale and interest in IPM; but this is an area where rice sales are important for most households, contributing an average of 43 per cent of cash income. It should not be surprising that the ability to apply a new technology (of any kind) to a commercial opportunity provides a strong incentive for uptake, and it is thus important to question the notion that LEIT is necessarily particularly suited to subsistence agriculture.

The patterns noted here are a reminder of the difficulties of providing useful technologies for farm production that does not make significant contributions to household income. There is some evidence that the availability of productive new technology allowed farmers to use LEIT to expand their commercial farming. This is particularly true in Honduras, where it appears that specific practices such as in-row tillage (as well as the experience of participating in local technology testing and generation) were a stimulus for some farmers to begin commercial vegetable cultivation. In the high-potential zone in Kenya, the option of growing grass strips for erosion control was made more attractive because of concomitant promotion of zero grazing. Farmers could harvest the Napier grass for their own cattle or sell it to neighbours.

A category as broad and open as LEIT challenges any attempts to draw general conclusions; but it is important that we try to be as precise as possible about what the literature and the case studies tell us. As a first step, it is possible to conclude that there is nothing about LEIT *per se* that guarantees that it will be preferentially adopted by resource-poor farmers or even that significant numbers of resource-poor farmers will find it useful. The three case studies all include examples in which resource-poor farmers were able to take advantage of LEIT; but in no case did the technology favour the less rather than the better resourced, nor was it particularly effective at reaching those at the bottom of the ladder. Most of the farmers in the case study communities were relatively poor, including those who took advantage of LEIT, and the innovations certainly contributed to improving their productivity and livelihoods. The projects have reason to be proud of their accomplishments. But the notion that a class of technology such as LEIT is, on its own, an effective tool for reaching the poorer sectors of farming communities is exceptionally simplistic and must be challenged.

It may be argued that in most instances the profile of an 'adopter' of a particular type of LEIT will not differ substantially from the profile emerging from an adoption study of an external input such as fertilizer. In neither case would we expect sharp differences, with farmers divided into easily distinguishable classes based on resources, practices and adoption. There are many factors that determine farmers' utilization of technology, and the fact that these relationships are often not simple or obvious supports a small industry of adoption studies and econometric analyses. But many studies find correlations between access to resources (land, capital and labour) and the uptake of innovations. Our case studies and review of the literature show that these patterns are equally evident when we examine the uptake of LEIT.

This is not to say that the particular examples of LEIT were not appropriate for the areas in which the projects were conducted. On the contrary, these techniques and ideas were taken up because the project management had sufficient experience, skill and flexibility to understand the production constraints and to work with farmers in developing useful interventions. There is no indication that the projects would have been better off promoting technologies based on purchased inputs. If such projects had been working in more favoured areas, their choice of technology would certainly have been different; but whether or not it would have been based on external inputs depends upon individual circumstances, not upon whether the area was classified as marginal or the farmers were classified as poor.

Labour and LEIT

Any examination of LEIT must be particularly concerned with the labour implications of the technology. We have already discussed the importance of challenging the notion that LEIT is necessarily labour-intensive; but we have also acknowledged that there are numerous instances (for instance, alley cropping and pest counting in IPM) where LEIT has been rejected by farmers because of high labour requirements. Labour is a fundamental determinant of a technology's acceptability to farmers. Moreover, the labour component cannot simply be assessed in terms of hours per hectare. The timing of the labour input during the season, the skill requirements (including the possibilities of learning to manage a technology more efficiently) and the difference between one-off investments (e.g. to establish a terrace) and annual requirements must all be taken into account.

The case studies provide examples of how labour requirements affect the utilization of various examples of LEIT. They also illustrate how off-farm labour opportunities influence investments in LEIT and underline the exceptional importance of hired labour.

In Kenya, the adoption of SWC techniques required an initial labour investment for establishment. Various types of vegetative and unploughed strips were the most common intervention, and these had to be laid out on the contour. The project staff helped with the establishment and the members of the catchment committee invested time in meetings and establishing contacts with participants. In a few cases, SWC structures that included several farms (such as drainage ditches) involved communal labour parties. Nevertheless, households in the high-potential area that were judged to be labour-deficient were less likely to establish SWC structures. The Kenyan farmers' choice of conservation technologies tended towards those with lower labour requirements (and those that offered additional economic advantages, such as strips of fodder grass). In general, farmers established only a few strips in their fields, at distances much greater than those recommended for effective erosion control.

In Honduras, in-row tillage adoption usually required considerable labour at the outset (although it is also possible to establish these structures gradually). About half of the labour for establishing these mini-terraces was hired. However, the study indicates that once in-row tillage is established it actually saves labour. This was the view of the majority of the adopters, and although there were a number of farmers who tried in-row tillage and abandoned it, very few of these cited labour requirements as a reason for losing interest in the technology.

In Sri Lanka, the major strategy for IPM was simply developing the confidence to reduce dependence upon insecticides and to recognize that many pest outbreaks did not call for an automatic application of chemicals. Any increase in time for monitoring fields would appear to be minimal, and no additional techniques or investments were required. It was thus a win–win situation, where lower labour (and/or cash) investment did not jeopardize yields. The other major example of LEIT that was promoted by the FFS (and by other extension activities) was straw incorporation. This required about five extra person days of labour per hectare; but it would appear that it offered significant yield gains, thus repaying the investment.

A factor that can affect the decision to invest additional labour in new technology is the competition with off-farm opportunities. Many farm families deploy a significant amount of labour in non-agricultural enterprises and this limits their ability to invest labour on the farm. In the Honduras case, there was no clear relationship between the level of off-farm labour and the use of in-row tillage (the fact that the technology apparently involves high initial labour but is subsequently labour-saving may cloud any relationship). In Kenya, we have seen that the adoption of SWC was much more likely for farmers with high crop incomes (and less likely for those whose income was derived more from off-farm activities). In Sri Lanka, those farmers with additional income from civil service jobs or trade were equally interested in techniques for lowering insecticide use; but those who derived a significant part of their income as day labourers tended to depend more heavily upon chemical insect control. The evidence would indicate that this latter group could have profited from reducing insecticide use (and saving time and labour); but it may be that the initial time investment of FFS participation to learn about these options was an unacceptable burden.

Any discussion of the labour component of LEIT needs to acknowledge the exceptional importance of hired labour. The idea that labour on small farms is necessarily provided by the household must be abandoned (along with the notion that resource-poor farmers are somehow 'rich' in labour). The three case studies deliberately focused on areas dominated by small farms producing crops for subsistence; but even in these environments the reliance on hired labour is substantial. Nearly half of the farms in the Honduras sample employed hired labour for some operations. More than half of the farms in the Kenya sample hired labour for weeding (and those farms have a higher rate of SWC uptake). The vast majority of the Sri Lankan farms employed hired labour for planting and harvesting rice (on holdings averaging less than 1ha), and the majority of those who applied insecticide did so by employing labourers.

The decision to utilize technologies (LEIT or otherwise) that require additional labour depends not only upon strategies for deploying household labour but also upon the availability of cash for hiring labour. Labour must thus be considered, in part, as an 'external input', further blurring the distinction between LEIT and other innovations. In a significant number of cases, the likelihood of utilizing LEIT will depend upon the availability of cash to hire the necessary labour. If the cash is generated from agricultural earnings, then the determinants of decisions on technology utilization may be fairly straightforward. This probably represents the majority of the cases in Sri Lanka, where hired labour is a key input for rice farms, the production of which contributes a significant proportion of farm household earnings. There are certainly similar examples in the Kenya and Honduras cases, where commercial agriculture (e.g. vegetables in Honduras) provides sufficient earnings for hiring farm labour. But if cash for hiring farm labour has to be generated through off-farm income, we need to ask if the pursuit of off-farm income allows enough time for learning new techniques or supervising labour, and whether farmers prefer to channel any earnings from off-farm labour to strengthen their livelihood diversification further or to invest in improved farming practices. In the Kenya case, it appears that those who earn substantial income from trade or businesses are less likely to invest in LEIT, while in Sri Lanka even those who earn most of their cash from trading or salaried jobs find it worthwhile to invest in rice farming innovations.

It is also important to consider the source of the hired labour and its implications for poverty reduction. In Sri Lanka, most of the hired labour is supplied by landless rural households (or farmers without access to irrigated land), and any extra income they can earn is welcome. In some cases, however, the poorest rice farming households also supply hired farm labour, and we have seen that these farmers are less likely to have the time to invest in learning about resource-conserving (and cost-saving) technology, which further jeopardizes the efficiency and competitiveness of their already marginal holdings. Similar questions can be asked in Kenya and perhaps Honduras, where farm labour has a higher probability of being drawn from landholding, rather than landless, households. Productive, labour-intensive technology can play a significant role in addressing rural poverty by providing additional employment; but if the poorest farming households have to earn part of their income working on others' fields, such technology may, in some cases, contribute to the demise of the most marginal holdings.

Labour is an exceptionally important element for understanding the future of LEIT. On the one hand, the case studies show that LEIT is not necessarily labour-intensive and that it may even be labour-saving. Even when LEIT involves an additional labour investment, farmers are willing to learn about and apply the new techniques if the gains in productivity are adequate. On the other hand, the examples show that farmers limit their investment of labour (hired or on-farm) and that the more labour-intensive LEIT options are less likely to attract interest. Because a growing amount of farm labour is hired rather than from the household, even among resource-poor farmers, it is difficult to distinguish LEIT from other innovations with respect to the use of on-farm resources and more difficult to argue that it is necessarily appropriate for, or targeted to, the poorest farming households.

Knowledge and LEIT

LEIT is also often characterized as knowledge-intensive; but the use of the term begs a number of questions. Knowledge may be associated with formal education, special training, farming experience, or indigenous skills and traditions. In addition, the knowledge requirements for managing a new technology may be addressed through a brief introduction or may imply a fairly extensive apprenticeship, and may even require continual upgrading. As was the case with labour, LEIT's knowledge requirements cover a very wide range of possibilities, making it difficult to offer any summary judgements on this type of technology.

In none of the case study examples was formal education associated with the ability to utilize LEIT, which should not be surprising (especially in view of the fairly narrow variability in formal schooling in each of the samples). Special training featured in all of the examples, but to a varying extent. Probably the least onerous requirements were found in the Kenya case, where attendance at a few meetings and participation in the establishment of SWC structures were all that was required. Farmers then learned to manage and adapt things such as vegetative strips in the course of their normal farming. In Honduras as well, farmers could be introduced to techniques such as in-row tillage or the use of cover crops in a fairly short time; but the techniques were sufficiently novel for participation in the project's on-farm experiments and demonstrations to help farmers gain confidence in the efficacy of these techniques. The most straightforward intervention, simply reducing the use of pesticides as a result of IPM training in Sri Lanka, seems, paradoxically, to have required an extensive amount of training. The FFS took up half a day each week during the crop cycle and although it featured a wide and varied curriculum, it would appear that one of its principal features was to provide farmers with the rationale and confidence to reduce their dependence upon insecticide (by acquainting them with relevant agro-ecological relationships), rather than teaching them new management skills. Thus, the time that may need to be spent in introducing a particular technology (and the time that farmers are required to invest) may be related less to the complexity of the technology itself and have more to do with providing farmers with new ways of looking at crop management or building up confidence to take new steps.

We must recognize that gaining knowledge takes time, and that, unlike providing extra labour, this additional time cannot be hired. Even when a LEIT innovation reduces labour requirements, farmers need to find the time (in the present) to learn about how to save time (in the future). It is therefore important that the most effective means for generating and transmitting the relevant knowledge are employed. Group methods were used to one degree or another in the case study examples, and they would seem to be an effective strategy. The group methods served variable purposes. In Kenya, the formation of a catchment committee facilitated communication with the project technicians, and group activities helped to validate the promotion of the SWC campaign and, at times, to enlist labour. In Honduras, farmers were encouraged to engage in on-farm experiments to test new technologies, and group activities facilitated social learning that helped farmers to share experiences and gain confidence. Social learning is a particularly important component of the FFS methodology and it would appear to be a key to changing farmers' attitudes towards insecticide use. In many cases, the utilization of LEIT also requires farmers to test and adapt the techniques, and this testing process provides the opportunity to share ideas with others. One problem is that group activity takes time, and those farmers with time to invest in group membership are not necessarily the poorest members of the farming community.

There is justification for attention to the knowledge requirements of LEIT, and although formal training or education is not necessarily a factor, farmers who have the time and interest to devote to learning new techniques or principles are more likely to take advantage of LEIT. In addition, farmers require time and initiative to experiment with and adapt some types of LEIT. However, it is difficult to make the case that LEIT is special in this regard, or that knowledge intensity is a defining characteristic of LEIT. For instance, the participants in the FFS in Sri Lanka came to appreciate the importance of straw management (an example of LEIT) and the management and use of single-nutrient fertilizers (an external input) for their farms. It is difficult to say which of these is more knowledge-intensive or whose mastery provides a greater degree of empowerment.

The emergence of a 'LEIT farmer'

LEIT is often promoted not as a single technology or even as a set of innovations, but rather as a fairly radical shift in attitude towards farming. In these cases, LEIT is seen as a way of helping farmers become more self-reliant (by lowering dependence upon external input markets) and as a way of stimulating stronger environmental awareness in farm management. The three case studies allow for an assessment of the degree to which this particular vision of LEIT is valid.

If the uptake of LEIT involves a change in farm management's perception of lower use of external inputs in general, we would expect that those farmers who utilized one type of LEIT would be more likely to utilize other examples of LEIT and/or to lower their use of external inputs. One way of visualising this is shown in Figure 8.1. The horizontal axis distinguishes between farmers who have adopted a particular example of LEIT (on the left) and those who have not (on the right). The vertical axis measures the proportion of each class of farmer that uses a particular external input. The expectation is that the slope of the line connecting the two points should be positive; a lower proportion of those who use a particular LEIT would be likely to use external inputs. Figures 8.2, 8.3 and 8.4 present examples from each of the case studies.

In Honduras, farmers who use in-row tillage are more likely to use fertilizer on their food crops, contradicting the hypothesis. A plausible explanation is that in-row tillage makes fertilizer use more efficient (perhaps more surprising, there is no difference in the amount of fertilizer used by adopters and non-adopters of cover crops, even though the innovation contributes to soil fertility, and in-row tillage users are no more likely than others to use organic fertilizers). A second example (that supports the hypothesis) is that farmers planting cover crops are less likely to use herbicide. In Kenya, there is almost no difference in the use of purchased seed between adopters and non-adopters of SWC. On the other hand, those who utilize SWC structures are more likely

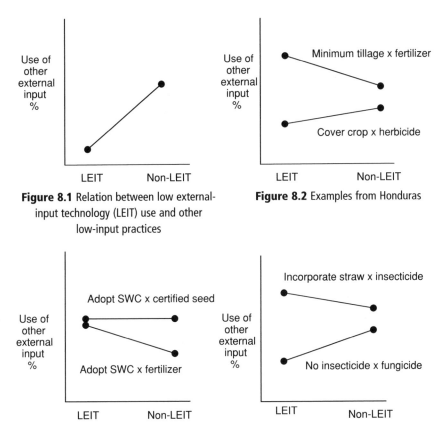

Figure 8.1 Relation between low external-input technology (LEIT) use and other low-input practices

Figure 8.2 Examples from Honduras

Figure 8.3 Examples from Kenya

Figure 8.4 Examples from Sri Lanka

to use fertilizer, almost certainly because the SWC provides an environment in which a profitable fertilizer response is more likely. In the Sri Lanka case, farmers who incorporate rice straw for soil fertility improvement use more insecticide than those who burn the straw. On the other hand, farmers who use no insecticide are also less likely to use fungicide than those farmers who make one or more insecticide applications.

These examples illustrate that there is no obvious tendency for the use of one type of LEIT to promote a general reduction in external inputs. In cases where a synergism is present, as where moisture- and soil-conserving practices make fertilizer use more profitable, 'low-input' soil management promotes 'high-input' soil fertility management. In other cases, there is a connection, as when the logic of reducing dependence upon chemical insecticides is extended to the use of fungicides. However, although Sri Lankan rice farmers who also grow vegetables would like to extend what they learned about IPM to these crops, they have not been able to do so. In most cases, farmers make decisions about various crop management technologies independently and pragmatically, without reference to any overarching philosophy regarding external inputs. In addition, farming systems evolve and require the consideration of new management practices. We have mentioned that the success of the LEIT project in Honduras is one of the factors that motivated some farmers to begin commercial vegetable production. This, in turn, has led to a significant increase in the use of pesticides (and there is little evidence that vegetable growers employing in-row tillage use less pesticide than their neighbours).

However, this is not to say that farmers are unaware of environmental concerns or that the LEIT projects were not successful in helping to stimulate farmers to consider the importance of resource conservation. In some cases, farmers credited environmental concerns for their interest in the LEIT; but there was no evidence that such a rationale took precedence over factors related to the economic returns and the costs related to the investment of labour and skills. Most Kenyan farmers are aware of the importance of soil conservation, a feature of many government and donor initiatives; but their investment in the technology is more closely linked to economic circumstances than to the expression of any environmental philosophy. They do, however, acknowledge that soil erosion is a concern for individual farms and do not expect that government should solve these problems for them. In Sri Lanka, FFS participants (and many non-participants) have reduced their use of insecticide, and many farmers also claim to use less fertilizer than previously. The participants are much more likely than the non-participants to cite environmental rather than economic reasons for these shifts; but there is little evidence that a lowering of insecticide use is linked to any other low-input practice, such as the use of organic soil amendments.

Farmers are certainly not blind to environmental concerns, but economic incentives remain the primary consideration in utilizing new technologies. Even though LEIT may be seen by some of its proponents as a holistic approach that can alter the way in which farmers look at the use of technology, the evidence to date indicates that farmers are much more eclectic and pragmatic in their choice of management practices than is assumed by the promotion of the image of a 'LEIT farmer'.

LEIT, experimentation and adaptation

Many examples of LEIT assume and encourage the adaptation of broad principles to individual farm circumstances. This may imply that much LEIT is less subject to simple imitation and requires more active engagement by the farmer with the new techniques. Such engagement may entail a learning cost; but the hope is that such costs will then be repaid by a strengthening of human capital in the direction of increased capacity for experimentation.

The successful utilization of LEIT often requires considerable adaptation and experimentation to accommodate principles and techniques to individual farm circumstances. This is true in the case study examples where, for instance, soil management techniques had to be adjusted to the codition of farmers' fields. In addition, LEIT projects often introduce the concept of formal experimentation. This was the case in Honduras, where farmers were encouraged to take part in and design their own experiments; it was also part of the FFS curriculum in Sri Lanka.

In Honduras, about one fifth of the participants who established in-row tillage made some type of modification subsequent to the initial adoption. Similarly, about one third of the Kenyan farmers who established an SWC structure made further adjustments, such as combining structures or altering the width or management of vegetative strips. In Sri Lanka, both participating and non-participating farmers could describe various local techniques that contributed to pest control; but the FFS participants did not lead the way in promoting or utilizing any novel pest-control techniques.

Farmers are constantly testing new ideas and making observations; but the prevalence of purposive, controlled comparisons in the field – what Sumberg and Okali (1997) call 'proactive' experiments – seems to be fairly low. Such experimental skills are difficult to define and assess; but the research uncovered relatively little evidence that the projects themselves led to any permanent increase in these skills. In each case, farmers (participants and non-participants) were asked to describe any crop management trials or small experiments that they had undertaken in recent years, particularly those that involved some type of comparison. The strongest evidence comes from Honduras, where the project put particular emphasis on developing experimental skills. About one fifth of the participants reported at least one experiment carried out subsequent to the project. In addition, farmers who utilized LEIT were more likely to try alternative pest control in maize (although they used no less chemical pesticide than other farmers). In the other two cases there was much less evidence of farmer experimentation.

The cases confirm that farmers are certainly capable of assessing and adapting technologies, and when LEIT options are presented in a participatory and egalitarian fashion, the incentives for such activities are greater. Projects that emphasize simple field experimentation (as in the Honduras case) can stimulate farmers to continue with their own investigations. But the case study examples do not support the view that LEIT projects, on their own, can set off a significant burst of local innovation.

Farmer-to-farmer dissemination of LEIT

Most LEIT projects represent a relatively high per-farmer investment, and the hope is often expressed that subsequent farmer-to-farmer diffusion can lower the costs of reaching the majority of farmers. Although most of the participants in the case study projects were enthusiastic about their experiences and reported describing ideas and concepts that they had learned to other farmers, there was only modest evidence that non-participants learned about particular techniques from project farmers. The results gave little hope that farmer-to-farmer diffusion would be sufficient to propel new techniques outward from the initial project investment. In Honduras, there was little adoption of in-row tillage and cover crops by non-participants, although about one quarter tried out the technologies. The majority who tried the techniques learned about them from various projects, although a significant number also reported learning from other farmers. In Kenya, farmers in some neighbouring communities established soil conservation techniques a few years after the project; the proximity of the project community was certainly important, though it must be remembered that there are various projects promoting these techniques in Kenya. In Sri Lanka, there was no evidence that FFS farmers played any significant role in persuading their neighbours to reduce insecticide use. Neighbour farmers were more likely to have remembered (and acted upon) information from the project participants about soil fertility management.

The lack of diffusion of LEIT technologies and principles outwards from these well-managed and relatively successful projects must be taken into account when considering future strategies for the promotion of LEIT. It would appear that in many cases farmers require a certain amount of hands-on experience before they are motivated to utilize many of these techniques, although there is considerable variation. In some cases (for example, vegetative strips or straw incorporation in rice fields) the techniques are quite visible to neighbouring farmers, who are capable of copying the ideas with a minimum of instruction. In other cases (for example, in-row tillage) farmers need some time to learn and experiment with the technique and the experience is more difficult to communicate. And in cases where complex principles are the basis for a change in practice (for example, agro-ecological analysis and the rationale for IPM), farmers may find it particularly challenging to articulate what they have learned to their neighbours.

Finally, it is important to consider the incentives for communication. On the one hand, the farmers in the case studies were generally quite willing to share what they had learned and even to proselytize, up to a point. In some cases, farmer leaders emerged or acquired additional skills and voice. But the idea that a LEIT project experience could be the basis of significant uncompensated farmer-to-farmer extension is not borne out by the evidence.

LEIT and social capital

There is much in the case studies that supports an emphasis on the development and support of farmer groups. Groups are often an efficient strategy for various extension activities. In addition, the case studies have confirmed the value of social learning, facilitated by farmer group activity. However, the undeniable value of farmer organization has led some supporters of LEIT to claim that successful group experiences (as part of a specific project in technology generation and design) often resulted in these groups staying together after the project and taking on other activities. This hypothesis deserves scrutiny.

There was very little evidence that participation in a LEIT project (and taking up useful new methods and practices) was sufficient to maintain the original groups formed for the projects or to lead to further organizational innovation. In Honduras, farmers who participated in the project belonged to more organizations; but the project did not lead to any further group initiatives. On the other hand, the satisfaction some project participants experienced in acquiring useful technology may have encouraged their subsequent participation in local organizations. The density of rural organization in Kenya is the highest of the three cases, and although some of the catchment project participants went on to participate in other activities, it is difficult to attribute this to the project itself. The catchment committees were a particular feature of this project, and they had considerable responsibilities in liaising with extension staff, organizing local activities and monitoring progress. Nevertheless, in only one site did part of a catchment committee remain together for further activities. In Sri Lanka, the FFS functioned for a season and then disbanded; the only exception was one case in which another extension activity (in local seed production) drew on the members of a previous FFS.

LEIT projects are established in communities where social networks are in place and there is some (although often minimal) experience of group activity. If the projects are competently managed, they can contribute to further group-led initiatives. But the idea that a technology generation project has a high probability of leading to further spontaneous group action is erroneous. Community organizations develop for various reasons (often in response to a succession of donor projects); but permanent, autonomous capacity requires particular incentives. The case of Guacamayas in Honduras is one of the exceptions that prove the rule. The utilization of LEIT in Guacamayas is very high and there is strong community organization. But pre-project activity had led to the development of a producers' co-operative, and the project built upon this organization and contributed new ideas. The community's location close to markets, its strong leaders and opportunities for producing organic coffee all contributed to further success. The LEIT project played an important role in this; but we can expect such a confluence of favourable circumstances in only a minority of cases.

Summary of case study experience

LEIT includes an exceptionally wide array of technologies, and the types of projects and strategies to promote LEIT range across the board in terms of comprehensiveness and competence. Nevertheless, this review of three large and well-managed projects, combined with an extensive review of the literature, provides a basis from which to offer some fairly general, but hopefully robust, conclusions. The research has demonstrated that there are many instances where farmers are able to take advantage of LEIT. Many of these technologies help farmers to become more productive and conserve resources, and these techniques deserve further investment in research, development and promotion. However, the use of LEIT in response to formal promotion efforts is much less widespread than some observers would hope. Rather than base our analysis on descriptions of individual farmers' success, or on estimates generated by the projects themselves, we made an independent assessment of the extent and nature of technology uptake, and the results are fairly sobering. Even when

LEIT is developed and adapted through well-managed projects, the actual uptake is often limited to a minority of participants and may feature only partial utilization of a particular technology.

LEIT does not necessarily behave in the way that either its supporters or sceptics maintain. On the one hand, LEIT is certainly not the rudimentary, hopelessly labour-demanding technology that some critics assume. There are many types of LEIT that are compatible with various farm labour profiles, and the technology, far from being primitive, often requires exceptional technical competence to develop and put in a form that farmers can then utilize and adapt. On the other hand, despite its focus on self-sufficiency and its concern for marginal farming conditions, most LEIT is not particularly targeted to the poorer members of farming communities; its adoption exhibits patterns similar to those of 'conventional' agricultural technology. In addition, the diffusion of the technology from farmer to farmer is often slower than that of other agricultural technologies. Finally, despite the fact that LEIT is often promoted as an approach to farming rather than as a particular technology, there are very few examples where LEIT efforts have served as a spark to ignite further individual innovation or group action.

The case studies found nothing to support the view that LEIT should be seen as the basis for a separate type of agricultural development strategy. Much LEIT is certainly appropriate for small farms; but it does not, on its own, address the major problems experienced by the large underclass of exceptionally poor smallholders who populate many rural communities. To promote the idea that a broad class of technology is specifically targeted to the poorest, or that particular types of technology generation and promotion will provide the inspiration for a rural renaissance, simply diverts attention from more comprehensive attempts to address rural poverty.

Technology and agricultural development

In reviewing the impact of the Green Revolution on the distribution of rural incomes, Freebairn (1995, p277) is concerned that although a technological solution to rural development is politically attractive, it can lead to the misleading assumption that 'the struggles and dislocations of altering social relationships, landholding patterns, political power sharing, and other deeply entrenched arrangements can be avoided'. That argument is easy to appreciate when considering innovation which depends upon the provision of seeds, fertilizers and irrigation; but our examination of LEIT indicates its relevance for what seems, at first glance, to be a class of technology more compatible with the needs of the poor. The spread and impact of LEIT have so far been much more modest than those of the Green Revolution; but there are no major examples of its adoption favouring the excluded or redressing rural inequality. Agricultural technology development and diffusion are absolutely essential for rural poverty reduction, and technology needs to be screened and targeted so that it has the highest relevance to poorer members of the community. Nevertheless, it is misleading to contend that a particular type of technology can substitute for more fundamental social and political reform. As an influential early critique of the Green Revolution's failure to reach poorer farmers concludes, 'the reason lies not so much in inadequate technology as in inappropriate institutions and poor policy' (Griffin, 1974, p255).

There is the danger that many of those who dismissed the Green Revolution for attempting to address social and political problems with technological interventions now find themselves embracing another kind of technology – LEIT – as an answer to rural inequality. In his exceptionally comprehensive review of the diffusion of all types of innovations Rogers (1995, p433) finds a similar pattern (expressed in sociological jargon): 'The consequences of the diffusion of innovation usually widen the socio-economic gap between the audience segments previously high and low in socio-economic status.' Rogers points out that the pattern is not inevitable; but contravening it requires an understanding of the various factors that may be associated with the underlying inequality. In his analysis, the three things that distinguish the privileged groups are greater awareness, superior ability to exchange information and access to additional resources. Our analysis has shown that the incentives, organization and resources of relatively wealthier farmers often make them the more likely beneficiaries of LEIT. We also have evidence that this is not an ironclad rule; but a much more realistic and dedicated effort is required before LEIT (or any agricultural technology) is a major force in addressing rural inequality.

This review of the performance of LEIT has shown that, despite its emphasis on local resources and knowledge, it is not particularly effective at reaching the poorer farmers in a community. There are many examples of LEIT that have improved the productivity and resilience of farmers in difficult environments, and there are many additional opportunities for this kind of technology. But we require an unsentimental examination of the nature of farming communities. The poor are often less likely to be able to take advantage of innovations because of inadequate human and social capital, and there are ways of building up these capacities (including through the promotion of LEIT). However, the people with the fewest assets, who are in most need of productive innovation, are also those who have the least flexibility and time to learn new techniques. Difficult decisions need to be made about the most appropriate development investments, and a realistic assessment of the possible contribution of agriculture to household livelihood must be balanced against opportunities for exit paths and economic diversification (de Janvry and Sadoulet, 2000; Berdegué and Escobar, 2002).

Thus, we need to move well beyond the myth of the self-provisioning peasant household, as well as the myth that is coming to replace it – of the strategically diversified rural household. Rural livelihood diversification, as we have seen, has several faces. On the one hand, new economic opportunities can mean that even households with significant agricultural assets are able to diversify their portfolios and, if agricultural markets are favourable, use increased earnings and contacts from other spheres to further their farming productivity. In these cases, the diversity does not necessarily interfere with the ability to take advantage of new technology, including LEIT. On the other hand, much rural livelihood diversification is, in fact, a shifting series of coping strategies demanded by the absence of any secure economic opportunity, and the household's scramble to keep its head above water seriously constrains its ability to consider new technology. When such innovation demands 'only' their time to learn or their labour to apply, these households balance that investment against other opportunities, including wage labour on the farms that may be better able to utilize the new technology. The type of technology that is on offer, and the way in which farmers gain control of it, is not irrelevant to the outcome. But the promotion of any kind of technology (from LEIT to transgenic crops) must address the heterogeneity of rural assets and opportunities and accept the limitations and constraints on agricultural pathways out of rural poverty.

Assessing LEIT

It is legitimate to discuss LEIT as a broad class of agricultural technology, but it is difficult to maintain that LEIT represents a distinct rural development strategy. The analysis in this book has included a literature review and case studies on particular examples of this type of technology. It found a wide variation in performance; but there was no difficulty in identifying cases where LEIT made a significant contribution to improving farming practices and production. Much of the variation in performance could be attributed to factors that commonly explain adoption patterns for any agricultural technology, including the availability of information, market opportunities, and access to labour and other resources.

On the other hand, the analysis found very little evidence to indicate that the uptake of particular types of LEIT was responsible for any broader shift in attitudes or practices towards 'sustainable' agriculture, or that LEIT served as a stimulus for the further development of human or social capital. The belief that a type of technology could be a spark for a significant shift in the capacities and organization of farming communities is exaggerated. That belief expresses itself as a vague hope in many who are interested in LEIT, a conviction among fieldworkers dedicated to developing the technology, and absolute dogma among some NGOs and academics who make a career of promoting LEIT. It is important to challenge the belief, not in order to extinguish the hopes of those who work on LEIT but, on the contrary, to see that they are not linked to, and ultimately discredited by, unrealistic assumptions. There is too much hard work, ingenuity and progress associated with the development of technological alternatives related to LEIT to let them become the hostage of a minority of ideologues. The picturesque image of the self-sufficient peasant farm trivializes the work of people developing LEIT and causes those who are acquainted with the complexity of rural communities to doubt the seriousness of the endeavour.

The idea that particular types of technology will contribute to the development of sustainable social capital needs to be re-examined. Many of the group-based activities used to develop and introduce LEIT are very effective; but their success does not necessarily depend upon the type of technology in question. The idealization of self-sufficient farming is linked to a similar oversimplification regarding the nature of rural organizations. The paradox is that although groups are very useful for promoting LEIT, the technology itself does not seem to be a particularly effective basis for the development of sustainable groups. The situation is further confused by the fact that most of the groups in question are the creations of donor projects. A succession of projects in a region is likely to draw on previously developed resources. Projects 'cherry pick' the most active groups from previous activities, and it is difficult to separate the emergence of autonomous, sustainable organizational capacity from the cosy collaboration between projects looking for a focus and a rural elite who know how to exploit project resources. In addition, only a minority of group activity appears to be directed towards 'bridging' local organizations with external resources. A recent review on the subject of community-based development concludes: 'There is virtually no reliable evidence on community participation projects actually increasing a community's capacity for collective action. This is clearly an area for further research' (Mansuri and Rao, 2004).

The lack of information on the performance and sustainability of community-based efforts is typical of the dearth of analysis available for assessing investments in LEIT. Much LEIT activity is the product of individual donor projects. Most such projects are the subject of only cursory end-of-project reviews, and very few donors invest in any kind of quantitative assessment. Donors do not have the time or the incentives to engage in a more thorough review and do not feel the need to organize an independent assessment of their progress. In the few cases where thorough assessments (internal or external) are undertaken, there does not seem to be any effort on the part of donors to compare results or to learn from each other. Despite the huge investment in project planning, donors devote inadequate resources to assessing results, notwithstanding the recent push for impact studies (which are useful but self-contained justifications for an investment, rather than an attempt to learn broader principles). Donors appear to do little to strengthen their own human capital regarding the capacity to learn from project experience.

The absence of data on LEIT is reflected in the relatively modest breadth of the literature reviewed in Chapter 4. The contrast between the few adoption studies available for LEIT and the vast literature on the uptake of conventional technology is notable. When Freebairn (1995) did a meta-analysis on the effects of the Green Revolution, he had more than 300 studies published in a 20-year period to examine. There are several reasons for the difference. There are still relatively few efforts at promoting LEIT; hence, the range of possibilities for study is smaller. In addition, it is often more difficult to define and identify instances of LEIT adoption; it is easy to ask if a farmer is using a new variety, but less straightforward to assess the degree of uptake of a soil conservation practice. More effort at assessing changes in farming practice due to LEIT projects is required. The same can be said for the new learning approaches that are often promoted in connection with LEIT (Loevinsohn et al, 2002).

Finally, we may hypothesize that a combination of contradictory incentives conspires to discourage examination of LEIT uptake. On the one hand, part of mainstream agricultural science does not take LEIT seriously and, hence, sees little reason to examine it in detail. On the other hand, those who support LEIT are under no pressure (e.g. from donors) to back up their claims about the spread and impact of their work, and so continue to rely upon upbeat and idiosyncratic reports to demonstrate progress. A few of the more exceptional reports get repeated and, at times, exaggerated – so we have the impression of yields doubling or tripling, farmers returning to the land and the land returning to resilient ecological health. But when we look for the evidence on which such claims are made, we are usually disappointed. The result is that the proponents of sustainable agriculture can draw upon a wide range of independent quantitative studies of conventional technology to point to deficiencies and bias in the use of external inputs, while critics of LEIT point to the embarrassing oversimplification of the more extreme proponents in order to justify their lack of interest in even considering the alternatives.

One example of the relative paucity of data is the reporting of the results of farmer field school projects for IPM. A review commissioned to analyse available evidence on the outcomes of this strategy (which has been in place for over a decade) found 25 studies, 18 of which were internal project evaluations (van den Berg, 2004). It would be unfair to suggest that internal evaluations, written by project staff and generally reporting very positive results, are necessarily biased or unreliable; but the nature of the incentives certainly suggests that the assessment net must be cast more widely than is currently the case. An influential global study of the uptake of sustainable agriculture practices (Pretty and Hine, 2001) relies almost entirely upon self-reports of project staff. The analysis was careful to screen out those reports that seemed unrealistic or based on questionable assumptions; but the contrast between this meta-analysis of LEIT, relying largely upon internal assessment, and similar analyses of conventional technology, such as Freebairn's (1995) on the Green Revolution, based entirely on independent studies, is instructive.

Too much of what we know, or think we know, about LEIT comes in brief reports, little 'human interest' boxes in various publications and photos of smiling groups of farmers. LEIT is certainly not the only offender in the escalating battle to capture the public's and donors' attention for sound-bite reports on rural development triumphs. Indeed, given the preponderance of efforts in conventional technology, LEIT is nowhere near the head of the class in the school of development project reporting that favours purported quotes from grateful farmers that a particular technology has now 'made hunger a thing of the past'. The challenge here is a more general one: getting donors and their projects to invest less in annual reports and self-congratulatory promotion and to pay more attention to careful, long-term monitoring of the course of agricultural change.

The challenge for LEIT is even more difficult because of its claims for longer-term impact on human and social capital. This was, of course, the reason that the case studies for this book were chosen to offer an examination of situations five or more years after the termination of project activity. The fact that relatively little growth in social capital can be identified in these cases is certainly not the last word on the subject. But we must face the fact that virtually the only 'evidence' we have in the literature for significant and permanent increases in group activity as a result of LEIT comes from self-reporting by those associated with project activity. These reports tend to shift into the passive voice when describing organizational growth ('groups were formed'), making it difficult to understand the dynamics, support and participation leading to such group development. A much more fine-grained and independent analysis of the evolution of local institutions in response to donor project inputs is called for.

Projects and scaling up

Despite serious concerns about the way in which some LEIT activities are reported (and the way that most are assessed), there is no doubt that support for this kind of technology development needs to be sustained and increased. It is possible to dismiss the notion that LEIT represents some kind of separate technological realm that automatically brings with it altered incentives and attitudes for farming, while at the same time acknowledging the importance of developing a more thoughtful and varied use of local resources and knowledge. To insist that LEIT stands apart (as a minority of its practitioners do) is simply inefficient and unrealistic. The key word in most attempts to improve the relevance of crop management research is 'integrated' (integrated natural resource management, nutrient management, pest management, etc.), and such integration involves the use of the widest possible range of inputs and techniques consistent with improving agricultural productivity in an environmentally responsible manner. To rule out possible contributors on the basis of self-sufficiency ideologies that do not originate with farmers themselves is an unfair burden imposed by outsiders who do not have to face the consequences.

Many of the methods that have been developed in conjunction with LEIT efforts that emphasize farmer participation, adaptive testing, group action and social learning are of great value as well. They are certainly much more widely applicable than is currently appreciated.

How can the lessons, techniques and motivations that characterize the best work in LEIT be promoted? The great temptation in an analysis such as this is to suggest how projects can be improved. The book has examined three very competently executed LEIT projects, documented their success and pointed out deficiencies. It is fairly straightforward to suggest that individual projects should be more responsive, empowering, decentralized and participatory, and that project strategies should be policy linked, joined up and scaled out. The 'better project' avenue is broad and attractive. It accommodates research institutes, NGOs and others who are ready to elaborate on and improve their last project; it provides donors with clean, bite-size targets for investing their funds; and it provides analysts and commentators with many interesting possibilities for dissection and comparison. Projects are the preferred path for the development assistance industry.

Unfortunately, there are a number of arguments against the supposition that simply improving the design of projects is going to be sufficient to make a significant contribution to the development of LEIT (or, indeed, to any other crop management technology directed towards resource-poor farmers). The factors to be considered include the dispersed nature of LEIT projects, the temptation towards high-profile technology, the inability to move beyond the pilot stage, and the high costs per farmer.

Many LEIT projects are limited in breadth and focus. They typically concentrate on a few specific technologies and often cover a relatively limited number of communities. There is nothing wrong with working with a restricted technology set as long as it responds to farmers' priorities rather than project mandates, and as long as there is a capacity for evolution and building on progress achieved. There is also nothing intrinsically wrong with starting on a small scale as long as there is some conception of the next steps that are necessary and a willingness to collaborate with similar efforts. The latter is a particular problem with LEIT, and it is not uncommon to find several separate projects working on similar issues (e.g. soil fertility) in the same region with little or no communication, coordination or joint learning.

In addition, technology themes proliferate and farmers may come into contact (simultaneously or serially) with efforts in, for example, participatory plant breeding, group formation for fodder crop nurseries, and a farmer field school for IPM. As each of these efforts usually brings with it a particular approach to learning, the resulting cacophony may limit individual effectiveness and undermine the credibility of these innovative approaches with farmers (Loevinsohn et al, 2002).

Each LEIT project tends to have a separate focus and unique methodological preferences that militate against wider application. There is competition for donor funds, and there is a need to sell a distinct product. For instance, it is much more attractive for donors and project personnel to invest in a project that attempts to organize organic cotton production in a relatively few communities than it is to devote the same resources to help a larger number of cotton farmers learn how to reduce (but not necessarily eliminate) their use of pesticides.

Most LEIT projects seem unable to pass beyond the pilot stage. There is no better example than the experience with farmer field schools for IPM, which have been successfully introduced in at least half a dozen Asian countries but have reached only a small proportion of farmers. LEIT projects are necessarily time-bound and almost always focus on particular technologies, hampering the possibility of responding to evolving conditions in the target areas.

One of the principal reasons that even the most successful LEIT projects have little widespread impact is their high costs. The success stories usually involve highly committed field workers (and typically involve many of the most energetic and innovative farmers). The costs of logistics, training and support are considerable. Calculating or projecting the cost per farmer of a LEIT project is not yet an exact science since estimates may be based on different assumptions about fixed costs and various approaches to the value of resources provided by donors or governments, as well as different ways of estimating the benefits (Fleischer et al, 2002). For instance, there is a wide range in estimates of the costs of farmer field schools. A programme in Zanzibar cost about

US$110 per farmer (Bruin and Meerman, 2001); IPM efforts in Indonesia and the Philippines are estimated to cost US$62 and US$47 per farmer, respectively (Quizon et al, 2001); and van den Berg et al (2002) use an estimate of US$12 per farmer in calculating the returns to farmer field schools in Sri Lanka. For hillside projects in Central America, Bunch (2002) estimates that a successful project in Guatemala cost a total of US$50 per participating farmer over the six-year life of the project, while the equivalent figure for the project in Honduras was US$300.

There is no disputing that good-quality agricultural development projects cost money; but resources are limited and there is a great deal that needs to be done to move beyond the technology-focused, small-scale pilot project pattern. The most common response for LEIT is 'scaling up' (Gonsalves, 2001; Gündel et al, 2001; Franzel et al, 2001), but the concept raises as many questions as it answers. The widespread use of the term in the development literature can be traced to the shift during the past two decades from top-down approaches to bottom-up grassroots strategies, often organized by NGOs (Uvin and Miller, 1994). A predominant concern has been how to take advantage of the lessons and contributions of a specific project and contribute to more widespread and sustainable changes, and a convenient catch-phrase that has become a mandatory element in the concluding recommendations of virtually every project report is 'scaling up' (various authors make not altogether consistent distinctions between 'scaling up' and 'scaling out', and for this brief discussion we shall simply use 'scaling up'). The term is now widely used, as several recent World Bank documents attest (Binswanger and Aiyar, 2003; World Bank, 2003).

One of the major problems is that the term is used for so many different things that it loses any possible utility. A comprehensive review defines scaling up as simply 'To efficiently increase the socio-economic impact from a small to a large scale of coverage' and provides a series of checklists that are so broad that it is impossible to imagine any project planner even consulting them (World Bank, 2003). The lack of precision in the term is symptomatic of the lack of clarity in the conceptualization of most project activity. The emphasis is on the immediate project and its resources, and little thought is given to any kind of more comprehensive strategy. The donor's urgency in moving funds and the contracting agency's priority in meeting payrolls conspire to extinguish any thoughts on long-term implications until reports are written, additional project phases are requested and scaling up becomes the preferred catch-phrase.

A particularly thorough exposition of the various meanings of scaling up is provided by Uvin and Miller (1994). They distinguish four principal categories for the term, and it is useful to apply these to the case of LEIT projects.

Their first category of scaling up is 'quantitative', the focus of which may include the working area, base or budget of a particular project; the most common example is, perhaps, simple replication of pilot projects. One problem is that such replication requires planning ahead, which is uncommon. Farrington and Lobo (1997) describe an exception for a watershed development programme in India that was careful to develop political support at various levels and map out a strategy for growth. But even in these cases, the question of costs is paramount, and it is rare that such expansion achieves any significant economies of scale. Indeed, initial pilot projects are often carried out with exceptional care that may be difficult to maintain on a broader scale (e.g. Place et al, 2003). There are certainly instances where LEIT innovations spread spontaneously (e.g. the use of a new cover crop); but we have seen that, in many cases to date, success is only achieved by repeating fairly intensive investment.

The second category of scaling up is 'functional'. This involves an expansion in the number or types of activity that an NGO or other entity takes responsibility for. One application to LEIT projects is the possibility that a project emphasizing soil fertility scales up not by establishing more soil fertility sites, but by developing farmers' capacities to address a wider range of issues. There are, for instance, examples of farmer field schools that maintain a focus on IPM, but assume responsibility for a broader range of topics as well. There is the widespread feeling that, to be successful, LEIT projects must go beyond specific technologies:

> To achieve scale and to also ensure sustainability in programmes of technological enhancement, the strengthening of local capacities to innovate may often be just as or even more important than the technologies themselves (Gonsalves, 2001).

This is surely a sensible advance over the crude notion of replicating specific technology projects. As Graves et al (2004, p530) point out:

> It is perhaps more important that farmers have a good understanding of the principles involved in nutrient management, pest management and crop interactions, rather than a detailed knowledge of a particular technique.

But we have seen that there is relatively little evidence to date that individual projects are able to contribute significantly to building such a broad set of skills.

The third category is 'political'. In this version of scaling up, the focus moves from service delivery to empowerment that allows participants to exert pressure on various levels of government. With respect to LEIT projects, this is similar to the commonly stated aim of building social capital (e.g. independent organizations). The importance of the goal is unassailable; but we must ask whether LEIT provides a sufficient basis for such political ambitions. Holt-Giménez (2001) discusses the experience of the *Campesino a Campesino* movement in Central America which, despite being a large and successful union of farmers based on agro-ecological agriculture, has not been able to evolve into an effective social movement or to gain political influence. Donors have a poor record of supporting local groups; any individual success is usually met by a rush from various donors to link up with a winner, and the resulting windfall swamps the fragile organization's capacities and distorts its incentives (e.g. Pretty, 1995, p143).

The fourth category of scaling up is 'organizational' and involves the evolution of mechanisms through which local organizations become self-sustaining or receive specific government support. At this point there is very little that is relevant to current LEIT activities and one would have to ask on what basis a LEIT activity, on its own, would hope to reach this level. Perhaps an example is provided by the Landcare movement in the Philippines, where (project-initiated) farmer organizations based on soil and water conservation activities have been able to form alliances with other farmer groups and also obtain support from local government (which has funds earmarked for conservation activities) (Mercado et al, 2001). A feature of Uvin and Miller's (1994) classification of scaling up, focused as it is on NGO activity, is that governments appear only as targets of pressure or providers of funding. Some other discussions of scaling up see government action in a more collaborative light; the World Bank (2003) review, for instance, describes policy change as an example of 'vertical scaling up'. But the more common interpretation focuses exclusively on dynamics from the grassroots. In a review of several conservation farming projects in the Philippines, Cramb et al (2000, p926) question this single-mindedness and suggest that there is:

> ... a strategic role for the state in coordinating an ongoing programme (rather than a series of resource-intensive projects ...) for upland development and conservation. To the extent that the rhetoric of community-based sustainable development obscures this reality, it is an ideology which is inimical to the interests of marginalized farmers and resource users.

The way forward

The final pages of studies such as this are typically graced by suggestions for improving project performance and scaling up the experience. If our judgement is that such solid, stepwise suggestions are unlikely, then a proposal that the entire project approach should be re-examined risks being considered as simply otherworldly. But if three outstanding projects produce only modest long-term results, and a thorough review of the literature supports the conclusion that LEIT is not making anywhere near the contribution that some people claim or many hope, then this should be an indication that we cannot continue with business as usual.

The principal response proposed in the conclusion to this chapter is that we are asking the wrong questions. Rather than searching for successful LEIT projects and asking how to expand their *scale*, we should be asking how this type of experience can contribute to a broader *scope* of rural development activity. An effort to ensure that farmers are able to take advantage of resource-conserving, productive, locally adapted technology cannot be based on a strategy of hundreds of small, uncoordinated projects in the hope that some proportion will catch on or serve as a spark that ignites wider development movements. We need much more general strategies that take us back to basics that address, for instance, what type of support is required for adequate rural education systems, what can be realistically expected from state extension services, what type of agricultural research is appropriate, and how broad-based, effective farmer organizations can be promoted. The current project approach involves unrealistic assumptions about the possibilities of engineering local innovations and too little attention to building underlying capacity. Project activity is an inefficient way of making up for deficiencies in basic education, information, markets and access to political power.

These questions take us away from short-term projects and suggest that external support must be redirected towards long-term institutional strengthening. They suggest that donors have to do a much better job of following up on their investments and that the development assistance industry, in general, must be more willing to work behind the scenes building up local capacities rather than investing in one-off projects. The prospects for such a shift in the near future are admittedly remote because there are few incentives for anyone. Donors need attractive evidence on the impacts of their project investments; but this is often anecdotal or collected within a rigid project evaluation framework meant to satisfy bureaucratic requirements rather than to promote learning (or contribute to objective exchange of experience among donors). Donors are held accountable by their taxpayers and treasuries; but such scrutiny is most often met by relatively superficial story lines rather than by messy and detailed analysis. Most NGOs working on LEIT are acutely aware of the necessity of pleasing major donors (who supply a significant proportion of their funding). They must also satisfy the members of the public who make charitable contributions, usually in the hope of seeing fairly immediate impact and often with the expectation that the image of a self-sufficient peasantry can be validated. Objective and detailed analysis entails risks that NGOs are so far not obliged to take. As Scholte (2003) points out in a description of church-funded projects, supporters need to be continually reassured and to feel that they are 'doing good'. The church projects do not intentionally distort reality, but they recognize that they are in competition with others to capture funds. Similarly, although some of the most competent and imaginative work in LEIT is done by research institutes and universities, their incentives are for producing elegant results and publishable materials, rather than for longer-term capacity-building behind the scenes.

This pessimism from the supply side may be tempered by somewhat more optimism on the demand side. Effort must be shifted towards organizing and articulating farmer demand. There is good reason to believe that:

> ... there are solid grounds for regarding resource managers, even poor ones, as autonomous, responsible, experimental and, though risk-averse, also opportunistic. Constraints not ignorance deter poor households. It follows from such an optimistic interpretation that they don't need to be lectured, cajoled, pressured or motivated but offered choices of technology ... information, experience and an enabling economic environment that make the

effort worthwhile (Mortimore, cited in Sayer and Campbell, 2004, p41).

The challenge is therefore to explore how external support can foster more effective demand from farmers that will, in turn, exert pressure on governments, donors and NGOs for institutional development. Three of the most relevant issues with respect to support for LEIT would seem to be:

1   the prospects and limitations of group activity;
2   an expansion in the choice of technologies available; and
3   support for the process of agricultural innovation.

We have already seen that there are strong justifications for encouraging the emergence of viable groups. Groups offer efficiencies in information transmission, form a basis for social learning and can help to foster environmental consciousness. Much agricultural technology requires hands-on learning, and groups can facilitate access to these opportunities. In addition, group action is usually the most effective way for farmers to draw on external resources. Rather than serving as gatekeepers for access to agricultural research stations, projects should strive to make farmers understand that public research belongs to them and that visits to an experimental station to seek attention to pressing problems are a right and not a brokered privilege. Similarly, organizing groups to exert pressure for more effective public extension may have higher payoffs than group formation in response to a brief donor project.

The problem is that current project-driven group formation is based on much too narrow a base, sometimes the development of a single technology. Groups will only be sustainable if they address major problems faced by farmers. Access to technology may be one of these; but it is unlikely that technology generation, on its own, will be the basis of a significant growth in viable groups. The probability that specific technological issues (such as IPM and soil erosion control) would form such a basis is even less likely. Groups need to offer as many advantages to farmers as possible in order to elicit significant commitment and offer opportunities for varying levels of participation. The transaction costs of group formation are significant, and there is no sense in repeatedly making such investments for a series of short-term, isolated interests. The various external initiatives with resources and rationales for group formation need to join forces.

While supporting group development wherever possible, it is important to not lose sight of other opportunities. Group initiatives are undoubtedly the preferred strategy for political action that draws in additional resources; but group activities are not the only mode for developing and transmitting information. We have seen that group activities may exclude certain classes of farmer and that people can learn about innovations from individual examples or by observing neighbours, leaders or risk-takers. Experiential learning is as important as social learning. In an age of ever more sophisticated communication technology, farmers deserve a better and more varied choice of information sources. As with group formation, a comprehensive strategy is required. For instance, while it is unlikely that the provision of agricultural information will be enough to keep an FM radio station in business, integrated support that links this to other farmer priorities (such as local news and entertainment) could ensure the existence of an additional, sustainable source of ideas.

Support for the diversification of information sources is linked to the diversification of technology options. LEIT projects often emphasize the importance of a 'basket of choices'; but in most cases the basket needs to be much bigger. The effort that projects invest in acquainting farmers with one small set of options should be broadened. This can only be done by expanding coordination among various technology generation efforts, and taking better advantage of a (hopefully) increasingly diversified array of information channels. Rather than investing in specialized events and demonstrations, more effort is required to provide comprehensive, periodic events such as farm shows, where a wide range of options and advice (provided by a number of different endeavours) is on offer. The roles of public extension and rural education systems must be considered in this light.

Attention to sustainable agriculture has highlighted the dangers of the tendency for conventional agricultural technology to simplify farming systems and to rely upon a narrow range of options, lowering diversity and resilience. But efforts in LEIT are also subject to faddism and uniformity – witness the careers of technologies such as vetiver grass or alley cropping. There are very few species promoted as cover crops or green manures, and when one catches on, it can face some of the same pest problems of the uniform fields of the Green Revolution. Velvet bean (*Mucuna pruriens*) ended up being promoted by more than 50 NGOs in Central America and Mexico (Buckles, 1995), and its demise in coastal Honduras is, in part, due to its inability to withstand an invasive weed (Neill and Lee, 2001). Methodological uniformity is also a problem, as in the rush to organize farmer field schools for almost every subject imaginable.

Simply recommending that options must be expanded is, of course, not a solution. The environmental and economic constraints under which most farmers operate admit little flexibility in resolving production problems, and there are a limited number of innovations that will be useful. Developing those innovations requires investment and commitment. New technologies often require champions who are willing to pursue controversial ideas, and these champions require resources to make their case. Rarely are technologies so immediately attractive that they spread without a significant period of demonstration, adaptation and debate. Thus, the options that are offered to farmers must be serious ones that have been through sufficient testing and assessment.

The development of adequate options raises the question of external support for innovation. Research and extension will be carried out by various public and civil society entities. This is not the place to discuss the importance of a more rational and coordinated approach to support for these activities; but it is relevant to remind ourselves that LEIT is not a simple seat-of-the-pants adaptation of traditional knowledge and that significant investment in good-quality research is required (e.g. Way and van Emden, 2000). Support for such research needs long-term planning. A recent review of natural resource management in the Consultative Group on International Agricultural Research (CGIAR) system recommends better coordination and devolution to (and support for) research in regional and national institutes (Lele, 2004). Coordination does not imply uniformity, and competition among approaches should be encouraged as long as there is an adequate system for assessing the results, supporting the successes and learning lessons from the less successful efforts. NGOs can offer support for farmer organization, advocacy and local technology testing and adaptation; but we have seen many examples that remind us of the importance of technical competence. Something as complex as LEIT is not effectively promoted by well-meaning but inexperienced generalists.

The innovations described under the rubric of LEIT offer important contributions for making agriculture more productive, protecting the environment and empowering farmers. But the current approach of pursuing myriad individual, unconnected projects is myopic and counterproductive. For those who believe that LEIT is the basis for a revolution in farming and rural life, the pursuit of these ideals through solitary projects seems to be an inescapable fate. But for the majority who see LEIT as an important contributor to a more general strengthening of smallholder farming, there is a responsibility to think more broadly. Funds, dedication and ingenuity need to be transferred from isolated, technology-specific efforts to building local capacities.

Those local capacities – among farmers, civil society, and public and market institutions – will encourage a shift towards more demand-driven innovation. With more choice and more information, farmers will be able to exact more accountability from those who offer technological alternatives. Donor strategies will not change overnight; but the exercise of better judgement in coordinating support for agricultural technology development, and much better and more transparent follow-up and assessment, will cost donors little, while making a tremendous contribution to the efficiency of external assistance. The major responsibility lies with national governments, who must foster strong institutions that support the interchange of information and experience. Those institutions (civil society groups, co-operatives, businesses, public education, and research and extension) have a better chance of development in a diversifying economy that allows economies of scope for institutional growth. The demand for technology development expands as stronger national economies provide increasing opportunities for the agricultural sector.

To say that LEIT must be the product of demand-driven technological change is not to abandon efforts at supporting technology generation. Rather, it is to recognize that sustainable agriculture can only be developed through sustainable local institutions and that the development of those institutions must be addressed in a systematic fashion.

References

Adégbidi, A., E. Gandonou and R. Oostendorp (2004) 'Measuring the productivity from indigenous soil and water conservation technologies with household fixed effects: A case study of hilly mountainous areas of Benin', *Economic Development and Cultural Change*, vol 52, pp313–346

Agrisystems (EA) Ltd (1998) *Impact Assessment Study – National Soil and Water Conservation Programme*, Nairobi, Ministry of Agriculture, Soil and Water Conservation Branch

Ali, M. (1999) 'Evaluation of green manure technology in tropical lowland rice systems', *Field Crops Research*, vol 61, pp61–78

Altieri, M. (1995) *Agroecology: The Science of Sustainable Agriculture* (second edition), Boulder, CO, Westview Press

Amanor, K. S. (1991) 'Managing the fallow: Weeding technology and environmental knowledge in the Krobo District of Ghana', *Agriculture and Human Values*, vol 8, pp5–13

Amanor, K. S. (1994) *The New Frontier: Farmers' Response to Land Degradation*, London, Zed Books

Anderson, S., S. Gündel and B. Pound with B. Triomphe (2001) *Cover Crops in Smallholder Agriculture: Lessons from Latin America*, London, ITDG Publishing

Arrellanes, P. G. (1994) 'Factors Influencing the Adoption of Hillside Agriculture Technologies in Honduras', MA thesis, Cornell University, Ithaca, NY

Arrellanes, P. G. and D. R. Lee (2001) 'The determinants of adoption of sustainable agriculture technologies in Honduras', Unpublished manuscript, Department of Applied Economics and Management, Cornell University, Ithaca, NY

Atampugre, N. (1993) *Behind the Lines of Stone: The Social Impact of a Soil and Water Conversation Project in the Sahel*, Oxford, Oxfam

Baland, J.-M. and J.-P. Platteau (1996) *Halting Degradation of Natural Resources*, Oxford, FAO and Clarendon Press

Barbier, E. (1990) 'The farm-level economics of soil conservation: The uplands of Java', *Land Economics*, vol 66, pp199–211

Barrett, C., J. Lynam, F. Place, T. Reardon and A. Aboud (2002) 'Towards improved natural resource management in African agriculture', in C. B.Barrett, F. Place and A. A. Aboud (eds) *Natural Resources Management in African Agriculture*, Wallingford, UK, CABI

Bassett, T. J. (1988) 'Breaking up the bottlenecks in food crop and cotton cultivation in northern Côte d'Ivoire', *Africa*, vol 58, pp147–173

Bebbington, A. (1992) *Searching for an 'Indigenous' Agricultural Development: Indian Organizations and NGOs in the Central Andes of Ecuador*, Working Paper No 45, Cambridge, Centre of Latin American Studies, University of Cambridge

Bebbington, A. (1999) 'Capitals and capabilities: A framework for analyzing peasant viability, rural livelihoods and poverty', *World Development*, vol 27, pp2011–2044

Bebbington, A., H. Carrasco, L. Peralbo, G. Ramón, J. Trujillo and V. Torres (1993) 'Fragile lands, fragile organizations: Indian organizations and the politics of sustainability in Ecuador', *Transactions of the Institute of British Geographers* N.S., vol 18, pp179–196

Bebbington, A. and T. Perreault (1999) 'Social capital, development and access to resources in highland Ecuador', *Economic Geography*, vol 75, pp395–418

Belknap, J. and W. E. Saupe (1988) 'Farm family resources and the adoption of noplow tillage in southwestern Wisconsin', *North Central Journal of Agricultural Economics*, vol 10, no 1, pp13–23

Bentley, J. W. (1989) 'What farmers don't know can't help them: The strengths and weaknesses of indigenous technical knowledge in Honduras', *Agriculture and Human Values*, vol 6, pp25–31

Berdegué, J. and G. Escobar (2002) *Rural Diversity, Agricultural Innovation Policies and Poverty Reduction*, AgREN Paper 122, London, ODI

Besley, T. and A. Case (1993) 'Modeling technology adoption in developing countries', *American Economic Review*, vol 83, pp396–402

Biggs, S. (1989) *Resource-Poor Farmer Participation in Research: A Synthesis of Experiences from Nine National Agricultural Research Systems*, OFCOR Comparative Study Paper No 3, The Hague, ISNAR

Bindlish, V. and R. Evenson (1997) 'The impact of T&V extension in Africa: The experience of Kenya and Burkina Faso', *World Bank Research Observer*, vol 12, pp183–201

Binswanger, H. and S. Aiyar (2003) *Scaling Up Community-Driven Development*, World Bank Policy Research Working Paper 3039, Washington, DC, World Bank

Birch-Thomsen, T., P. Fredriksen and H.-O. Sano (2001) 'A livelihood perspective on natural resource management and environmental change in semiarid Tanzania', *Economic Geography*, vol 77, pp41–66

Birkhaeuser, D., R. Evenson and G. Feder (1991) 'The economic impact of agricultural extension: A review', *Economic Development and Cultural Change*, vol 39, pp607–650

Blaikie, P., J. Cameron and D. Seddon (2002) 'Understanding 20 years of change in West-Central Nepal: Continuity and change in lives and ideas', *World Development*, vol 30, pp1255–1270

Böhringer, A., E. Ayuk, R. Katanga and S. Ruvuga (2003) 'Farmer nurseries as a catalyst for developing sustainable land use systems in southern Africa. Part A: Nursery productivity and organization', *Agricultural Systems*, vol 77, pp187–201

Boyd, C. and T. Slaymaker (2000) *Re-Examining the 'More People Less Erosion' Hypothesis: Special Case or Wider Trend?* Natural Resource Perspectives No 63, London, ODI

Branford, S. and J. Rocha (2002) *Cutting the Wire: The Story of the Landless Movement in Brazil*, London, Latin American Bureau

Braun, A. R., G. Thiele and M. Fernández (2000) *Farmer Field Schools and Local Agricultural Research Committees: Complementary Platforms for Integrated Decision-Making in Sustainable Agriculture*, AgREN Paper No 105, London, ODI

Bray, F. (1986) *The Rice Economies*, Oxford, Basil Blackwell

Brown, D. and C. Korte (1997) *Institutional Development of Local Organisations in the Context of Farmer-Led Extension: The Agroforestry Programme of the Mag'uugmad Foundation*, AgREN Paper No 68, London, ODI

Brown, L. D. and D. Ashman (1996) 'Participation, social capital and intersectoral problem solving: African and Asian cases', *World Development*, vol 24, pp1467–1479

Brown, J. S. and P. Duguid (2000) *The Social Life of Information*, Cambridge, MA, Harvard Business School Press

Bruin, G. and F. Meerman (2001) *New Ways of Developing Agricultural Technologies: The Zanzibar Experience with Participatory Integrated Pest Management*, Wageningen, The Netherlands, Wageningen University and Research Centre

Bryceson, D., C. Kay and J. Mooij (eds) (2000) *Disappearing Peasantries? Rural Labour in Africa, Asia and Latin America*, London, Intermediate Technology Publications

Buckles, D. (1995) 'Velvetbean: A "new" plant with a history', *Economic Botany*, vol 49, no 1, pp13–25

Buckles, D., B. Triomphe and G. Sain (1998) *Cover Crops in Hillside Agriculture: Farmer Innovation with Mucuna*, Ottawa, Canada, and Mexico, DF, IDRC and CIMMYT

Bunch, R. (1982) *Two Ears of Corn: A Guide to People-Centered Agricultural Improvement*, Oklahoma City, OK, World Neighbors

Bunch, R. (1999) 'Learning how to "Make the Soil Grow": Three case studies on soil recuperation adoption and adaptation from Honduras and Guatemala', in M. McDonald and K. Brown (eds) *Issues and Options in the Design of Soil and Water Conservation Projects*, Proceedings of a workshop held in Llandudno, Conwy, UK, 1–3 February

Bunch, R. (2002) 'Changing productivity through agroecological approaches in Central America: Experiences from hillside agriculture', in N. Uphoff (ed) *Agroecological Innovations*, London, Earthscan

Burton, M., D. Rigby and T. Young (1999) 'Analysis of the determinants of adoption of organic horticultural techniques in the UK', *Journal of Agricultural Economics*, vol 50, pp47–63

Buttel, F., G. Gillespie Jr., R. Janke, B. Caldwell and M. Sarrantonio (1986) 'Reduced-input agricultural systems: A critique', *The Rural Sociologist*, vol 6, no 5, pp350–370

Butterworth, J., B. Adolph and B. S. Reddy (2003) *How Farmers Manage Soil Fertility: A Guide to Support Innovation and Livelihoods*, Hyderabad, Andhra Pradesh Rural Livelihoods Programme, and Chatham, UK, NRI

Byerlee, D. (1998) 'Knowledge-intensive crop management technologies: Concepts, impacts and prospects in Asian agriculture', in P. Pingali and M. Hossain (eds) *Impact of Rice Research*, Manila, IRRI

Cameron, L. (1999) 'The importance of learning in the adoption of high-yielding variety seed', *American Journal of Agricultural Economics*, vol 81, pp83–94

Carson, R. (1962) *Silent Spring*, Boston, MA, Houghton Mifflin

Carter, J. (1995) *Alley Farming: Have Resource-Poor Farmers Benefited?* Natural Resource Perspectives No 3, London, ODI

Cary, J. and R. Wilkinson (1997) 'Perceived profitability and farmers' conservation behaviour', *Journal of Agricultural Economics*, vol 48, pp13–21

Cassman, K.G. (1999) 'Ecological intensification of cereal production systems: Yield potential, soil quality and precision agriculture', *Proceedings of the National Academy of Science (USA)*, vol 96, pp5952–5959

Central Bank of Sri Lanka (2002) *Annual Report*, Colombo, Central Bank of Sri Lanka

Chaker, M. with H. Abbassi and A. Laouina (1996) 'Mountains, foothills and plains: Investing in SWC in Morocco', in C. Reij, I. Scoones and C. Toulmin (eds) *Sustaining the Soil: Indigenous Soil and Water Conservation in Africa*, London, Earthscan

Chambers, R. (1997) *Whose Reality Counts?* London, Intermediate Technology Publications

Chambers, R., A. Pacey and L.A. Thrupp (eds) (1989) *Farmer First: Farmer Innovation and Agricultural Research*, London, Intermediate Technology Publications

Charles, D. (2001) *Lords of the Harvest*, Cambridge, MA, Perseus

Chinnappa, B. N. and W. P. T. Silva (1977) 'Impact of the cultivation of high-yielding varieties of paddy on income and employment', in B. H. Farmer (ed) *Green Revolution? Technology and Change in Rice-growing Areas of Tamil Nadu and Sri Lanka*, London, Macmillan

Choi, H. and C. Coughenour (1979) 'Socioeconomic aspects of no-tillage agriculture: A case study of farmers in Christian County, Kentucky', Department of Sociology, University of Kentucky, Kentucky

Christian Aid (1999) *Selling Suicide: Farming, False Promises and Genetic Engineering in Developing Countries*, London, Christian Aid

CIP–UPWARD (Users' Perspectives with Agricultural Research and Development) (2003) *Farmer Field Schools: From IPM to Platforms for Learning and Empowerment*, Los Baños, the Philippines, UPWARD

Coleman, J. S. (1988) 'Social capital in the creation of human capital', *American Journal of Sociology*, vol 94 (supplement), ppS95–S120

Collinson, M. (1983) *Farm Management in Peasant Agriculture*, Boulder, CO, Westview Press

Collinson, M. (ed) (2000) *A History of Farming Systems Research*, Wallingford, UK, CABI and Rome, FAO

Conklin, H. C. (1957) *Hanunoo Agriculture: A Report on an Integral System of Shifting Cultivation in the Philippines*, Rome, FAO

Conley, T. and C. Udry (2001) 'Social learning through networks: The adoption of new agricultural technologies in Ghana', *American Journal of Agricultural Economics*, vol 83, pp668–673

Conway, G. (1997) *The Doubly Green Revolution*, Harmondsworth, Penguin

Conway, G. and J. Pretty (1991) *Unwelcome Harvest*, London, Earthscan

Cooper, P., R. Leakey, M. Rao and L. Reynolds (1996) 'Agroforestry and the mitigation of land degradation in the humid and sub-humid tropics of Africa', *Experimental Agriculture*, vol 32, pp235–290

Cowan, R. and P. Gunby (1996) 'Sprayed to death: Path dependence, lock-in and pest control strategies', *The Economic Journal*, vol 106, pp521–542

Cramb, R. A., J. N. M. Garcia, R. V. Gerrits and G. C. Saguiguit (2000) 'Conservation farming projects in the Philippine uplands: Rhetoric and reality', *World Development*, vol 28, pp911–927

Crowley, E. and S. Carter (2000) 'Agrarian change and the changing relationships between toil and soil in Maragoli, Western Kenya (1900–1994)', *Human Ecology*, vol 28, pp383–414

Daberkow, S. and W. McBride (2003) 'Farm and operator characteristics affecting the awareness and adoption of precision agriculture technologies in the US', *Precision Agriculture*, vol 4, pp163–177

David, C. and K. Otsuka (eds) (1994) *Modern Rice Technology and Income Distribution in Asia*, Boulder, CO, Lynne Rienner

Defoer, T. and A. Budelman (eds) (2000) *Managing Soil Fertility in the Tropics: A Resource Guide for Participatory Learning and Action Research*, Amsterdam, Royal Tropical Institute

Dembélé, I., L. Kater, D. Koné, Y. Koné, B. Ly and A. Macinanke (2001) 'Seizing new opportunities: Soil-fertility management and diverse livelihoods in Mali', in I. Scoones (ed) *Dynamics and Diversity: Soil Fertility and Farming Livelihoods in Africa*, London, Earthscan

de Jager, A. (1991) 'Towards self-experimenting village groups', in B. Haverkort, J. van den Kamp and A. Waters-Bayer (eds) *Joining Farmers' Experiments*, London, Intermediate Technology Publications

de Jager, A., D. Onduru and C. Walaga (2004) 'Facilitated learning in soil fertility management: assessing potentials of low-external-input technologies in East African farming systems', *Agricultural Systems*, vol 79, pp205–223

de Janvry, A. and E. Sadoulet (2000) 'Rural poverty in Latin America: Determinants and exit paths', *Food Policy*, vol 25, pp389–409

Dilts, R. (2001) 'Scaling up the IPM movement', *LEISA Magazine*, vol 17, no 3, pp18–20

Doberman, A. (2004) 'A critical assessment of the system of rice intensification (SRI)', *Agricultural Systems*, vol 79, pp261–281

Douglass, G. (1984) 'The meanings of agricultural sustainability' in G. Douglass (ed) *Agricultural Sustainability in a Changing World Order*, Boulder, CO, Westview Press

Douthwaite, B., V. Manyong, J. Keatinge and J. Chianu (2002) 'The adoption of alley farming and *Mucuna*: Lessons for research, development and extension', *Agroforestry Systems*, vol 56, pp193–202

Dvorák, K. A. (1996) *Adoption Potential of Alley Cropping*, Resource and Crop Management Research Monograph No 23, Ibadan, IITA

Ekboir, J. (ed) (2002) *CIMMYT 2000–2001 World Wheat Overview and Outlook: Developing No-Till Packages for Small-Scale Farmers*, Mexico, DF, CIMMYT

Ellis, E. and S. Wang (1997) 'Sustainable traditional agriculture in the Tai Lake region of China', *Agriculture, Ecosystems and Environment*, vol 61, pp177–193

Ellis, F. (1998) 'Household strategies and rural livelihood diversification', *Journal of Development Studies*, vol 35, pp1–38

Ellis, F. and N. Mdoe (2003) 'Livelihoods and rural poverty reduction in Tanzania', *World Development*, vol 31, pp1367–1384

Elvin, M. (1993) 'Three thousand years of unsustainable growth: China's environment from archaic times to present', *East Asian History*, vol 6, pp7–46

Erenstein, O. (1996) *Evaluating the Potential of Conservation Tillage in Maize-based Farming Systems in the Mexican Tropics*, NRG Reprint Series 96–01, Mexico, DF, CIMMYT

Fakih, M., T. Rahardjo and M. Pimbert (2003) *Community Integrated Pest Management in Indonesia*, London, IIED, and Brighton, IDS

Falck, M. and A. E. Díaz (1999) *Equidad y pobreza rural en Honduras: Realidad y propuestas*, Tegucigalpa, Escuela Agrícola Panamericana, Zamorano

Farrington, J. and A. Bebbington with K. Wellard and D. Lewis (1993) *Reluctant Partners? Non-Governmental Organizations, the State and Sustainable Agricultural Development*, London, Routledge

Farrington, J. and C. Lobo (1997) *Scaling up Participatory Watershed Development in India: Lessons from the Indo-German Watershed Development Programme*, Natural Resource Perspectives 17, London, ODI

Feder, G., R. E. Just and D. Zilberman (1985) 'Adoption of agricultural innovations in developing countries: A survey', *Economic Development and Cultural Change*, vol 33, pp255–298

Feder, G. and R. Slade (1984) 'The acquisition of information and the adoption of new technology', *American Journal of Agricultural Economics*, vol 66, pp312–320

Fernandez-Cornejo, J., C. Hendricks and A. Mishra (forthcoming) 'Technology adoption and off-farm household income: The case of herbicide-tolerant soybeans', *Journal of Agricultural and Applied Economics*

Fernandez-Cornejo, J. and W. McBride (2002) *Adoption of Bioengineeered Crops*, ERS Agricultural Economic Report No AER810, Washington, DC, USDA

Fine, B. (2001) *Social Capital versus Social Theory*, London, Routledge

Fischer, A. J., A. J. Arnold and M. Gibbs (1996) 'Information and the speed of innovation adoption', *American Journal of Agricultural Economics*, vol 78, pp1073–1081

Fleischer, G., H. Waibel and G. Walter-Echols (2002) 'Transforming top-down agricultural extension to a participatory system: A study of costs and prospective benefits in Egypt', *Public Administration and Development*, vol 22, pp309–322

Foster, A. D. and M. R. Rosenzweig (1995) 'Learning by doing and learning from others: Human capital and technical change in agriculture', *Journal of Political Economy*, vol 103, pp1176–1209

Fox, H. (1979) 'Local farmers' associations and the circulation of agricultural information in nineteenth-century England', in H. Fox and R. Butlin (eds) *Change in the Countryside: Essays on Rural England 1500–1900*, Institute of British Geographers Special Publication No 10, London, Institute of British Geographers

Francis, C. (ed) (1986) *Multiple Cropping Systems*, New York, John Wiley and Sons

Francis, P. (1987) 'Land tenure systems and agricultural innovations: The case of alley farming in Nigeria', *Land Use Policy*, vol 4, pp305–319

Franzel, S. (1999) 'Socioeconomic factors affecting the adoption potential of improved tree fallows in Africa', *Agroforestry Systems*, vol 47, pp305–321

Franzel, S., P. Cooper and G. L. Denning (2001) 'Scaling up the benefits of agro-forestry research: Lessons learned and research challenges', *Development in Practice*, vol 11, pp524–534

Franzel, S., D. Phiri and F. Kwesiga (2002) 'Assessing the adoption potential of improved fallows in eastern Zambia', in S. Franzel and S. J. Scherr (eds) *Trees on the Farm: Assessing the Adoption Potential of Agroforestry Practices in Africa*, Wallingford, UK, CABI

Freebairn, D. (1995) 'Did the Green Revolution concentrate incomes? A quantitative study of research reports', *World Development*, vol 23, pp265–279

Fujisaka, S. (1993) 'A case of farmer adaptation and adoption of contour hedgerows for soil conservation', *Experimental Agriculture*, vol 29, pp97–105

Gambetta, D. (ed) (1988) *Trust: Making and Breaking Cooperative Relations*, Oxford, Basil Blackwell

Gautam, M. (2000) *Agricultural Extension: The Kenya Experience*, Washington, DC, World Bank

Geertz, C. (1971) *Agricultural Involution: The Processes of Ecological Change in Indonesia*, Berkeley, CA, University of California Press

Giller, K. (2002) 'Targeting management of organic resources and mineral fertilizers: Can we match scientists' fantasies with farmers' realities?', in B. Vanlauwe, J. Diels, N. Sanginga and R. Merckx (eds) *Integrated Plant Nutrient Management in Sub-Saharan Africa: From Concept to Practice*, Wallingford, UK, CABI

Gonsalves, J. F. (2001) 'Going to scale. What we have garnered from recent work-shops', *LEISA Magazine*, vol 17, no 3, pp6–10

Granovetter, M. S. (1972) 'The strength of weak ties', *American Journal of Sociology*, vol 78, no 6, pp1360–1380

Graves, A., R. Matthews and K. Waldie (2004) 'Low external input technologies for livelihood improvement in subsistence agriculture', *Advances in Agronomy*, vol 82, pp473–555

Grieshop, J., F. Zalom and G. Miyao (1988) 'Adoption and diffusion of Integrated Pest Management innovations in agriculture', *Bulletin of the Entomological Society of America*, vol 34, no 2, pp72–78

Griffin, K. (1974) *The Political Economy of Agrarian Change*, London, Macmillan

Grindle, M. S. (1988) *Searching for Rural Development: Labor Migration and Employment in Mexico*, Ithaca, NY, Cornell University Press

Grootaert, C., G.-T. Oh and A. Swamy (2002) 'Social capital, household welfare and poverty in Burkina Faso', *Journal of African Economies*, vol 11, pp4–38

Grootaert, C. and T. van Bastelaer (2002) 'Introduction and overview', in C. Grootaert and T. van Bastelaer (eds) *The Role of Social Capital in Development*, Cambridge, Cambridge University Press

Gugerty, M. K. and M. Kremer (2002) 'The impact of development assistance on social capital: Evidence from Kenya', in C. Grootaert and T. van Bastelaer (eds) *The Role of Social Capital in Development*, Cambridge, Cambridge University Press

Gündel, S., J. Hancock and S. Anderson (2001) *Scaling-up Strategies for Research in Natural Resources Management: A Comparative Review*, Chatham, UK, NRI

Haggblade, S. and G. Tembo (2003) *Conservation Farming in Zambia*, EPTD Discussion Paper No 108, Washington, DC, IFPRI

Haggblade, S., G. Tembo and C. Donavan (2004) *Household Level Financial Incentives to Adoption of Conservation Agricultural Technologies in Africa*, Food Security Research Project Working Paper No 9, East Lansing, MI, Michigan State University

Harding, D., J. K. Kiara, and K. Thomson (1996) *Soil and Water Conservation in Kenya: The Development of the Catchment Approach and Structured Participation – Led by the Soil and Water Conservation Branch of the Ministry of Agriculture of Agriculture, Kenya*, Paris, OECD, Dossier prepared for the OECD/DAC Workshop on Capacity Development in the Environment, 4–6 December 1996, Rome

Harrison, E. (1996) 'Digging fish ponds: Perspectives on motivation in Luapula Province, Zambia', *Human Organization*, vol 55, pp270–278

Harrison, P. (1987) *The Greening of Africa*, London, Paladin

Harriss, J. (2001) *Depoliticizing Development: The World Bank and Social Capital*, New Delhi, LeftWord Books

Harriss, J. and P. de Renzio (1997) '"Missing link" or analytically missing? The concept of social capital', *Journal of International Development*, vol 9, pp919–937

Harriss-White, B. (2003) *India Working: Essays on Society and Economy*, Cambridge, Cambridge University Press

Hassane, A. (1996) 'Improved traditional planting pits in the Tahoua Department, Niger', in C. Reij, I. Scoones and C. Toulmin (eds) *Sustaining the Soil: Indigenous Soil and Water Conservation in Africa*, London, Earthscan

Hassane, A., P. Martin and C. Reij (2000) *Water Harvesting, Land Rehabilitation and Household Food Security in Niger*, Rome, IFAD

Hayami, Y. and V. W. Ruttan (1985) *Agricultural Development: An International Perspective* (second edition), Baltimore, MD, Johns Hopkins University Press

Hazell, P. B. R. and C. Ramasamy (1991) *The Green Revolution Reconsidered*, Baltimore, MD, Johns Hopkins University Press

Heinrich, J. (2001) 'Cultural transmission and the diffusion of innovation: Adoption dynamics indicate that biased cultural transmission is the predominate force in behavioral change', *American Anthropologist*, vol 103, pp992–1013

Hellin, J. and K. Schrader (2003) 'The case against direct incentives and the search for alternative approaches to better land management in Central America', *Agriculture, Ecosystems and the Environment*, vol 99, pp61–81

Heong, K. and M. Escalada (1999) 'Quantifying rice farmers' pest management decisions: Beliefs and subjective norms in stem borer control', *Crop Protection*, vol 18, pp315–322

Heong, K., M. Escalada, N. Huan and V. Mai (1998) 'Use of communication media in changing farmers' pest management in the Mekong Delta, Vietnam', *Crop Protection*, vol 17, pp413–425

Hesse-Rodríguez, M. (1994) *Sembradores de Esperanza: Conservar para Cultivar y Vivir*, Tegucigalpa, PROCONDEMA – Guaymuras – Comunica

Hogg, R. (1988) 'Water harvesting and agricultural production in semi-arid Kenya', *Development and Change*, vol 19, pp69–87

Holmberg, G. (1985) *An Economic Evaluation of Soil Conservation in Kalia Sub-location, Kitui District*, Nairobi, Ministry of Agriculture and Livestock Development, Soil and Water Conservation Division

Holt-Giménez, E. (2001) 'Scaling up sustainable agriculture. Lessons from the Campesino a Campesino movement', *LEISA Magazine*, vol 17, no 3, pp27–29

Howard-Borjas, P. and K. Jansen (2000) 'Ensuring the future of sustainable agriculture: What could this mean for jobs and livelihood in the 21st century?', Paper presented at the conference The Role of the Village in the 21st century: Crops, Jobs and Livelihood, 15–17 August, Hannover, Germany

Humphries, S., J. Gonzales, J. Jimenez and F. Sierra (1999) *Searching for Sustainable Land Use Practices in Honduras: Lessons from a Program of Participatory Research with Hillside Farmers*, AgREN Paper 104, London, ODI

Jansen, H. G., A. Damon, J. Pender, W. Wielmaker and R. Schipper (2003) *Policies for Sustainable Development in the Hillsides of Honduras: A Quantitative*

Livelihoods Approach, Washington, DC, IFPRI

Jeger, M. (2000) 'Bottlenecks in IPM', *Crop Protection*, vol 19, pp787–792

Johnson, D. and L. Lewis (1995) *Land Degradation: Creation and Destruction*, Oxford, Blackwell

Jones, K. (2002) 'Integrated pest and crop management in Sri Lanka', in N. Uphoff (ed) *Agroecological Innovations*, London, Earthscan

Jones, S. (2002) 'A framework for understanding on-farm environmental degradation and constraints to the adoption of soil conservation measures: Case studies from highland Tanzania and Thailand', *World Development*, vol 30, pp1607–1620

Kaboré, D. and C. Reij (2004) *The Emergence and Spreading of an Improved Traditional Soil and Water Conservation Practice in Burkina Faso*, EPTD Discussion Paper No 114, Washington, DC, IFPRI

Keeley, J. and I. Scoones (2000) 'Knowledge, power and politics: The environmental policy-making process in Ethiopia', *Journal of Modern African Studies*, vol 38, pp89–120

Kelly, V., M. Syalla, M. Galiba and D. Weight (2002) 'Synergies between natural resource management practices and fertilizer technologies: Lessons from Mali', in C. B. Barrett, F. Place and A. A. Aboud (eds) *Natural Resources Management in African Agriculture*, Wallingford, UK, CABI

Kenmore, P. (1996) 'Integrated pest management in rice', in G. Persley (ed) *Biotechnology and Integrated Pest Management*, Wallingford, UK, CABI

Kerr, J. (2002) 'Watershed development, environmental services and poverty alleviation in India', *World Development*, vol 30, pp1387–1400

Kerr, J. and S. Kolavalli (1999) *Impact of Agricultural Research on Poverty Alleviation: Conceptual Framework with Illustrations from the Literature*, EPTD Discussion Paper No 56, Washington, DC, IFPRI

Khan, Z., J. Pickett, L. Wadhams and F. Muyekho (2001) 'Habitat management strategies for the control of cereal stemborers and striga in maize in Kenya', *Insect Science Applications*, vol 21, no 4, pp375–380

Kimani, M., N. Mihindo and S. Williamson (2000) 'We too are proud to be researchers', in G. Stoll (ed) *Natural Crop Protection in the Tropics*, Weikersheim, Margraf Verlag

Konde, A., D. Dea, E. Jonfa, F. Folla, I. Scoones, K. Kena, T. Berhanu and W. Tessema (2001) 'Creating gardens: The dynamics of soil–fertility management in Wolayta, southern Ethiopia', in , in I. Scoones (ed) *Dynamics and Diversity: Soil Fertility and Farming Livelihoods in Africa*, London, Earthscan

Krishna, A. (2002) *Active Social Capital*, New York, Columbia University Press

Kuyvenhoven, A. (2004) 'Creating an enabling environment: policy conditions for less-favored areas', *Food Policy*, vol 29, pp407–429

Landers, J. N. (2001) *Zero Tillage Development in Tropical Brazil*, FAO Agricultural Services Bulletin 147, Rome, FAO

Leach, M. and R. Mearns (eds) (1996) *The Lie of the Land*, Oxford, James Currey

Leaf, M. J. (1984) *Song of Hope: The Green Revolution in a Panjab Village*, New Brunswick, NJ, Rutgers University Press

LEISA Magazine (2003) 'Learning with Farmer Field Schools', *LEISA Magazine,* special issue, vol 9, no 1

Lele, U. (2004) *The CGIAR at 31: An Independent Meta-Evaluation of the Consultative Group on International Agricultural Research*, Washington, DC, World Bank

Lightfoot, C., K. Dear and R. Mead (1987) 'Intercropping sorghum with cowpea in dryland farming systems in Botswana. II. Comparative stability of alternative cropping systems', *Experimental Agriculture*, vol 23, pp435–442

Lin, N. (2001) *Social Capital: A Theory of Social Structure and Action*, Cambridge, Cambridge University Press

Lipton, M. with R. Longhurst (1989) *New Seeds and Poor People*, London, Unwin Hyman

Little, P. D. and D. W. Brokensha (1987) 'Local institutions, tenure and resource management in East Africa', in D. Anderson and R. Grove (eds) *Conservation in Africa*, Cambridge, Cambridge University Press

Lockeretz, W. (1990) 'What have we learned about who conserves soil?', *Journal of Soil and Water Conservation*, vol 45, pp517–523

Lockeretz, W. (1991) 'Information requirements of reduced-chemical production methods', *American Journal of Alternative Agriculture*, vol 6, pp97–103

Lockeretz, W. and S. Wernick (1980) 'Commercial organic farming in the corn belt in comparison to conventional practices', *Rural Sociology*, vol 45, pp708–722

Loevinsohn, M., J. Berdegué and I. Guijit (2002) 'Deepening the basis of rural resource management: Learning processes and decision support', *Agricultural Systems*, vol 73, pp3–22

Long, N. (1968) *Social Change and the Individual*, Manchester, Manchester University Press

Low, A. (1986) *Agricultural Development in Southern Africa*, London, James Currey

Low, A. (1993) 'The low-input, sustainable agriculture (LISA) prescription: A bitter pill for farm-households in southern Africa', *Project Appraisal*, vol 8, no 2, pp97–101

Lumley, S. (1997) 'The environment and the ethics of discounting: An empirical analysis', *Ecological Economics*, vol 20, pp71–82

Lutz, E., S. Pagiola and C. Reiche (eds) (1994) *Economic and Institutional Analyses of Soil Conservation Projects in Central America and the Caribbean*, World Bank Environment Paper 8, Washington, DC, World Bank

Madeley, J. (2002) *Food for All*, London, Zed Books

Malley, Z., B. Kayombo, T. Willcocks and P. Mtakwa (2004) 'Ngoro: An indigenous, sustainable and profitable soil, water and nutrient conservation system in Tanzania for sloping land', *Soil and Tillage Research*, vol 77, pp47–58

Mansuri, G. and V. Rao (2004) 'Community-based and driven development: A critical review', *World Bank Research Observer*, vol 19, pp1–39

Manyong, V. M., V. A. Houndékon, P. C. Sanginga, P. Vissoh and A. N. Honlonkou (1999) *Mucuna Fallow Diffusion in Southern Benin*, Ibadan, Nigeria, IITA

Mayoux, L. (2001) 'Tackling the down side: Social capital, women's empowerment and micro-finance in Cameroon' *Development and Change*, vol 32, pp435–464

McDonald, M. and K. Brown (2000) 'Soil and water conservation projects and rural livelihoods: Options for design and research to enhance adoption and adaptation', *Land Degradation and Development*, vol 11, pp343–361

Meinzen-Dick, R., M. DiGregorio and N. McCarthy (2004) 'Methods for studying collective action in rural development', *Agricultural Systems*, vol 82, pp197–214

Mercado, A. R., M. Patindol and D. P. Garrity (2001) 'The Landcare experience in the Philippines: Technical and institutional innovations for conservation farming', *Development in Practice*, vol 11, pp495–508

Miclat-Teves, A. and D. Lewis (1993) 'Overview', in J. Farrington and D. Lewis (eds) *Non-Governmental Organizations and the State in Asia*, London, Routledge

Mitchell, J. C. (1973) 'Networks, norms and institutions', in J. Boissevain and J. C. Mitchell (eds) *Network Analysis: Studies in Human Interaction*, The Hague, Mouton

Moore, M. (1985) *The State and Peasant Politics in Sri Lanka*, Cambridge, Cambridge University Press

Morris, M. L., R. Tripp and A. A. Dankyi (1999) *Adoption and Impacts of Improved Maize Technology: A Case Study of the Ghana Grains Development Project*, Economics Program Paper 99–01, Mexico, DF, CIMMYT

Morse, S. and W. Buhler (1997) *Integrated Pest Management: Ideals and Realities in Developing Countries*, Boulder, CO, Lynne Rienner

Moser, C. and C. Barrett (2003) 'The disappointing adoption dynamics of a yield-increasing, low external-input technology: The case of SRI in Madagascar', *Agricultural Systems*, vol 76, pp1085–1100

Murray, G. F. (1994) 'Technoeconomic, organizational, and ideational factors as determinants of soil conservation in the Dominican Republic', in E. Lutz, S. Pagiola and C. Reiche (eds) *Economic and Institutional Analyses of Soil Conservation Projects in Central America and the Caribbean*, World Bank Environment Paper 8, Washington, DC, World Bank

Murton, J. (1999) 'Population growth and poverty in Machakos District, Kenya', *Geographical Journal*, vol 165, pp37–46

Murwira, K., H. Wedgewood, C. Watson and E. Win with C. Tawney (2000) *Beating Hunger: The Chivi Experience*, London, Intermediate Technology Publications

Nagy, J., J. Sanders and H. Ohm (1988) 'Cereal technology interventions for the West African semi-arid tropics', *Agricultural Economics*, vol 2, pp197–208

Narayan, D. (1997) *Voices of the Poor: Poverty and Social Capital in Tanzania*, Environmentally and Socially Sustainable Development Studies and Monographs Series 20, Washington, DC, World Bank

Narayan, D. and L. Pritchett (1997) *Cents and Sociability: Household Income and Social Capital in Rural Tanzania*, Policy Research Working Paper 1796, Washington, DC, World Bank, Social Development and Development Research Group

Naylor, R. (1994) 'Herbicide use in Asian rice production', *World Development*, vol 22, pp55–70

Neill, S. and D. Lee (2001) 'Explaining the adoption and disadoption of sustainable agriculture: The case of cover crops in Northern Honduras', *Economic Development and Cultural Change*, vol 49, pp793–820

Netting, R. McC. (1968) *Hill Farmers of Nigeria*, Seattle, WA, University of Washington Press

Noordin, Q., A. Niang, B. Jama and M. Nyasimi (2001) 'Scaling up adoption and impact of agroforestry technologies: Experiences from western Kenya', *Development in Practice*, vol 11, pp509–523

NRC (National Research Council) (1989) *Alternative Agriculture*, Washington, DC, National Academy Press

NRC (1993) *Vetiver Grass: A Thin Green Line Against Erosion*, Washington, DC, National Academy Press

Nugaliyadde, L., T. Hidaka and M. P. Dhanapala (1997) 'Pest management practices of rice farmers in Sri Lanka', in K. L. Heong and M. M. Escalada (eds) *Pest Management of Rice Farmers in Asia*, Los Baños, the Philippines, IRRI

Okoth, J., G. Khisa and T. Julianus (2003) 'Towards self–financed farmer field schools', *LEISA Magazine* vol 19(1) pp28–29

Orr, A. (2003) 'Integrated pest management for resource-poor African farmers: Is the emperor naked?', *World Development*, vol 31, pp 831–845

Orr, A., B. Mwale and D. Saiti (2001) 'What is an integrated pest management "strategy"? Explorations in southern Malawi', *Experimental Agriculture*, vol 37, pp473–494

Orton, L. (2003) *GM Crops – Going against the Grain*, London, ActionAid

Ostrom, E., L. Schroeder and S. Wynne (1993) *Institutional Incentives and Sustainable Development*, Boulder, CO, Westview Press

Ouedraogo, M. and V. Kaboré (1996) 'The zaï: A traditional technique for the rehabilitation of degraded land in the Yatenga, Burkina Faso', in C. Reij, I. Scoones and C. Toulmin (eds) *Sustaining the Soil: Indigenous Soil and Water Conservation in Africa*, London, Earthscan

Owens, M. and B. Simpson (2002) 'Farmer field schools as an extension strategy: A West African Experience', Paper prepared for Workshop on Extension and

Rural Development: A Convergence of Views on Institutional Approaches, World Bank, Washington, DC, 12–14 November

Palladino, P. (1996) *Entomology, Ecology and Agriculture: The Making of Scientific Careers in North America, 1885–1985*, Amsterdam, Harwood Academic Publishers

Parrott, N. and T. Marsden (2002) *The Real Green Revolution: Organic and Agroecological Farming in the South*, London, Greenpeace Environmental Trust

Pender, J. (1996) 'Discount rates and credit markets: Theory and evidence from rural India', *Journal of Development Economics*, vol 50, pp257–296

Pender, J. and J. Kerr (1998) 'Determinants of farmers' indigenous soil and water conservation investments in semi-arid India', *Agricultural Economics*, vol 19, pp113–125

Pender J. and S. Scherr (2002) 'Organizational development and natural resource management: Evidence from Central Honduras', in R. Meinzen-Dick (ed) *Innovation in Natural Resource Management: The Role of Property Rights and Collective Action in Developing Countries*, Baltimore, MD, and London, Johns Hopkins University Press

Pender, J., S. Scherr and G. Durón (2001) 'Pathways of development in the hillside areas of Honduras: Causes and implications for agricultural production, poverty and sustainable resource use', in D. R. Lee and C. B. Barrett (eds) *Tradeoffs or Synergies? Agricultural Intensification, Economic Development and the Environment*, Wallingford, UK, CABI

Pingali, P. L. and G. A. Carlson (1985) 'Human capital, adjustments in subjective probabilities and the demand for pest controls', *American Journal of Agricultural Economics*, vol 67, pp853–861

Piters, B. and P. Ndakidemi (2000) 'Mixed farming in northern Tanzania', in A. Budelman and T. Defoer (eds) *PLAR and Resource Flow Analysis in Practice*, Amsterdam, Royal Tropical Institute

Place, F., M. Adato, P. Hebinck and M. Omosa (2003) *The Impact of Agroforestry-Based Soil Fertility Replenishment Practices on the Poor in Western Kenya*, FCND Discussion Paper No 160, Washington, DC, IFPRI

Place, F., S. Franzel, J. DeWolf, R. Rommelse, F. Kwesiga, A. Niang and B. Jama (2002) 'Agroforesty for soil fertility replenishment: Evidence on adoption processes in Kenya and Zambia', in C. B. Barrett, F. Place and A. A. Aboud (eds) *Natural Resources Management in African Agriculture*, Wallingford, UK, CABI

Place, F., B. Swallow, J. Wangila and C. Barrett (2002) 'Lessons for natural resource management technology adoption and research', in C. B. Barrett, F. Place and A. A. Aboud (eds) *Natural Resources Management in African Agriculture*, Wallingford, UK, CABI

Ponnamperuma, F. N. (1984) 'Straw as a source of nutrients for wetland rice', in IRRI (ed) *Organic Matter and Rice*, Los Baños, the Philippines, IRRI

Pontius, J., R. Dilts and A. Bartlett (eds) (2002) *Ten Years of IPM Training in Asia – From Farmer Field School to Community IPM*, Bangkok, FAO

Pretty, J. (1991) 'Farmers' extension practice and technology adaptation: Agricultural revolution in 17th- to 19th-century Britain', *Agriculture and Human Values*, vol 8, pp132–148

Pretty, J. (1995) *Regenerating Agriculture*, London, Earthscan

Pretty, J. (2002) *Agri-Culture: Reconnecting People, Land and Nature*, London, Earthscan

Pretty, J. and R. Hine (2001) *Feeding the World with Sustainable Agriculture: A Summary of New Evidence*, SAFE–World Final Report, Colchester, UK, University of Essex

Pretty, J., J. Morrison and R. Hine (2003) 'Reducing food poverty by increasing agricultural sustainability in developing countries', *Agriculture, Ecosystems and Environment*, vol 95, no 1, pp217–234

Pretty J., J. Thompson and J. K. Kiara (1995) 'Agricultural regeneration in Kenya: The catchment approach to soil and water conservation', *Ambio*, vol 24, no 1, pp 7–15

Pretty, J. and H. Ward (2001) 'Social capital and the environment', *World Development*, vol 29, pp209–227

Price, L. (2001) 'Demystifying farmers' entomological and pest management knowledge: A methodology for assessing the impacts on knowledge from IPM–FFS and NES interventions', *Agriculture and Human Values*, vol 18, pp153–176

Putnam, R. D. (1993) *Making Democracy Work*, Princeton, NJ, Princeton University Press

Putnam, R. D. (2000) *Bowling Alone*, New York, Simon and Schuster

Quizon, J., G. Feder and R. Murgai (2001) 'Fiscal sustainability of agricultural extension: The case of the farmer field school approach', *Journal of International Agriculture and Extension Education*, Spring, pp13–23

Reardon, T., J. Taylor, K. Stamoulis, P. Lanjouw and A. Balisacan (2000) 'Effects of non-farm employment on rural income inequality in developing countries: An investment perspective', *Journal of Agricultural Economics*, vol 51, pp266–288

Reij, C., I. Scoones and C. Toulmin (eds) (1996) *Sustaining the Soil: Indigenous Soil and Water Conservation in Africa*, London, Earthscan

Reij, C. and A. Waters-Bayer (eds) (2001) *Farmer Innovation in Africa*, London, Earthscan

Reijntjes, C., B. Haverkort and A. Waters-Bayer (1992) *Farming for the Future*, Leusden, ILEIA

Reinhardt, N. (1983) 'Commercialization of agriculture and rural living standards: El Palmar, Colombia, 1960–1979', *Journal of Economic History*, vol 43, no 1, pp251–259

Renkow, M. (2000) 'Poverty, productivity and production environment: A review of the evidence', *Food Policy*, vol 25, pp463–478

Republic of Kenya (1996) *Programme Proposal 1997–2000: Soil and Water Conservation Programme (GOK/SIDA)*, Nairobi, Ministry of Agriculture, Livestock Division and Marketing, Agriculture Engineering Division, Soil and Water Conservation Branch

Ridgley, A.-M. and S. Brush (1992) 'Social factors and selective technology adoption: The case of integrated pest management', *Human Organization*, vol 51, pp367–378

Roberts, K. and J. Coutts (1997) *A Broader Approach to Common Resource Management: Landcare and Integrated Catchment Management in Queensland, Australia*, AgREN Paper No 70, London, ODI

Rogers, E. M. (1995) *Diffusion of Innovations* (fourth edition), New York, Free Press

Rola, A. C., S. B. Jamais and J. B. Quizon (2002) 'Do farmer field school graduates retain and share what they learn? An investigation in Iloilo, Philippines', *Journal of International Agriculture and Extension Education*, vol 9, no 1, pp65–76

Rola, A. C. and P. L. Pingali (1993) *Pesticides, Rice Productivity and Farmers' Health: An Economic Assessment*, Los Baños, the Philippines, IRRI

Röling, N. and E. van de Fliert (1994) 'Transforming extension for sustainable agriculture: The case of integrated pest management in rice in Indonesia', *Agriculture and Human Values*, vol 11, no 2/3, pp96–108

Ruben, R. (2001) 'Economic conditions for sustainable agriculture: A new role for the market and the state', *LEISA Magazine*, October, pp52–53

Ruben, R. and D. Lee (2000) 'Combining internal and external inputs for sustainable agricultural intensification', IFPRI Policy Brief 65, Washington, DC, IFPRI

Ruben, R. and J. Pender (2004) 'Rural diversity and heterogeneity in less-favoured areas: The quest for policy targeting', *Food Policy*, vol 29, pp303–320

Ruthenberg, H. (1976) *Farming Systems in the Tropics* (second edition), Oxford, Clarendon Press

Sain, G. and H. Barreto (1996) 'The adoption of soil conservation technology in El Salvador: Linking productivity and conservation', *Journal of Soil and Water Conservation*, vol 51, pp313–321

Sanders, R. (2000) *Prospects for Sustainable Development in the Chinese Countryside*, Aldershot, Ashgate

Sayer, J. and B. Campbell (2004) *The Science of Sustainable Development*, Cambridge, Cambridge University Press

Scarborough, V., S. Killough, D. Johnson and J. Farrington (1997) *Farmer-Led Extension: Concepts and Practices*, London, Intermediate Technology Publications

Scholte, E. (2003) 'The virtual reality of Protestant development aid in the Netherlands', in P. Quarles van Ufford and A. K. Giri (eds) *A Moral Critique of Development*, London, Routledge

Schultz, T. W. (1964) *Transforming Traditional Agriculture*, New Haven, CT, Yale University Press

Scoones, I. (ed) (2001) *Dynamics and Diversity: Soil Fertility and Farming Livelihoods in Africa*, London, Earthscan

Sen, A. (1999) *Development as Freedom*, Oxford, Oxford University Press

Settle, W., H. Araiwan, E. Astuti, W. Cahyana, A. Hakim, D. Hindayana, A. Lestari and Pajarningsih (1996) 'Managing tropical rice pests through conservation of generalist natural enemies and alternative prey', *Ecology*, vol 77, no 7, pp1975–1988

Seyoum, E., G. Battese and E. Fleming (1998) 'Technical efficiency and productivity of maize producers in eastern Ethiopia: A study of farmers within and outside the Sasakawa–Global 2000 project', *Agricultural Economics*, vol 19, pp341–348

Shiva, V. (1993) *Monocultures of the Mind: Perspectives on Biodiversity and Biotechnology*, London, Zed Books

Slingerland, M. and M. Masdewel (1996) 'Mulching on the Central Plateau of Burkina Faso', in C. Reij, I. Scoones and C. Toulmin (eds) *Sustaining the Soil: Indigenous Soil and Water Conservation in Africa*, London, Earthscan

Smale, M. and V. Ruttan (1997) 'Social capital and technical change: The *groupements naam* of Burkina Faso' in C. Clague (ed) *Institutions and Economic Development*. Baltimore, MD, Johns Hopkins University Press

Smil, V. (1991) 'Population growth and nitrogen: An exploration of a critical existential link', *Population and Development Review*, vol 17, no 4, pp569–601

Smil, V. (2000) *Feeding the World: A Challenge for the Twenty-First Century*, Cambridge, MA, MIT Press

Smith, K. R. (2002) 'Does off-farm work hinder "smart" farming?' *Agricultural Outlook,* September, pp28–30

Smucker, G., T. A. White and M. Bannister (2002) 'Land tenure and the adoption of agricultural technology in Haiti' in R. Meinzen-Dick, A. Knox, F. Place and B. Swallow (eds) *Innovation in Natural Resource Management*, Baltimore, MD, Johns Hopkins University Press

Soule, M. J. (1997) *Farmer Assessment of Velvetbean as a Green Manure in Veracruz, Mexico: Experimentation and Expected Profits,* Natural Resources Group Paper 97–02, Mexico, DF, CIMMYT

Spencer, J. (1990) *A Sinhala Village in a Time of Trouble*, Oxford, Oxford University Press

Stocking, M. (1996) 'Soil erosion: Breaking new ground', in M. Leach and R. Mearns (eds) *The Lie of the Land*, Oxford, James Currey

Stoll, S. (2002) *Larding the Lean Earth*, New York, Hill and Wang

Stone, G. D., R. McC Netting and M. P. Stone (1990) 'Seasonality, labor scheduling and agricultural intensification in the Nigerian Savannah', *American Anthropologist*, vol 92, pp7–23

Stoop, W. (2003) 'The system of rice intensification (SRI) from Madagascar: Myth or missed opportunity?', Unpublished report

Stoop, W., N. Uphoff and A. Kassam (2002) 'A review of agricultural research issues raised by the system of rice intensification (SRI) from Madagascar: Opportunities for improving farming systems for resource–poor farmers', *Agricultural Systems*, vol 71, pp249–274

Suazo, L. (2004). 'Agua, Pendiente, Líderes, Organización: Agricultura Sostenible en Guacamayas', Estudio de Caso del Proyecto LEIT de ODI/CIDICCO, mimeo

Sumberg, J. and C. Okali (1997) *Farmers' Experiments: Creating Local Knowledge*, Boulder, CO, Lynne Rienner

Thomas, D. B., E. K. Biamah, A. M. Kilewe, L. Lundgren and B. O. Mochoge (1989) *Soil and Water Conservation in Kenya: Proceedings of the Third National Workshop Kabete, Nairobi, 16–19 September 1986*, Nairobi, Department of Agricultural Engineering, University of Nairobi and Stockholm Swedish International Development Authority

Thomas, D. B., A. Ericksson, M. Grunder and J. K. Mburu (1997) *Soil and Water Conservation Manual for Kenya*, Nairobi, English Press Ltd for Soil and Water Conservation Branch, Ministry of Agriculture, Livestock Development and Marketing

Tiffen, M., M. Mortimore and F. Gichuki (1994) *More People, Less Erosion*, Chichester, John Wiley

Tisdell, C. (1988) 'Sustainable development: Differing perspectives of ecologists and economists, and relevance to LDCs', *World Development*, vol 16, pp373–384

Toulmin, C. and I. Scoones (2001) 'Ways forward? Technical choices, intervention strategies and policy options', in I. Scoones (ed) *Dynamics and Diversity: Soil Fertility and Farming Livelihoods in Africa*, London, Earthscan

Traoré, N., R. Landry and N. Amara (1998) 'On-farm adoption of conservation practices: The role of farm and farmer characteristics, perceptions, and health hazards', *Land Economics*, vol 74, pp114–127

Tripp, R. (1996) 'Biodiversity and modern crop varieties: sharpening the debate', *Agriculture and Human Values*, vol 13, no 4, pp48–63

Tripp, R. (2001) *Seed Provision and Agricultural Development*, Oxford, James Currey

Tripp, R. and A. Ali (2001) *Farmers' Access to Natural Pest Control Products: Experience from an IPM Project in India*, AgREN Paper No 113, London, ODI

Uphoff, N. (ed) *Agroecological Innovations*, London, Earthscan

Uphoff, N. (2002b) 'The agricultural development challenges we face', in N. Uphoff (ed) *Agroecological Innovations*, London, Earthscan

Uphoff, N. (2002c) 'Opportunities for raising yields by changing management practices: The system of rice intensification in Madagascar', in N. Uphoff (ed) (2002a) *Agroecological Innovations*, London, Earthscan

Uphoff, N. (2002d) 'Introduction', in N. Uphoff (ed) *Agroecological Innovations*, London, Earthscan

Uvin, P. and D. Miller (1994) *Scaling Up: Thinking Through the Issues*, World Hunger Program Research Report 1994–1, Providence, RI, Brown University

Vaessen, J. and J. de Groot (2004) *Evaluating Training Projects on Low External Input Agriculture: Lessons from Guatemala*, AgREN Paper No 139, London, ODI

van den Berg, H. (2004) *IPM Farmer Field Schools: A Synthesis of 25 Impact Evaluations*, Rome, Global IPM Facility

van den Berg, H., H. Senerath, and L. Amarasinghe (2002) *Participatory IPM in Sri Lanka: A broad-scale and an in-depth impact analysis*, Wageningen, The Netherlands, FAO Programme for Community IPM in Asia

van der Ploeg, J. (1990) *Labor, Markets and Agricultural Production*, Boulder, CO, Westview Press

Veron, R. (1999) *Real Markets and Environmental Change in Kerala, India*, Aldershot, Ashgate

Versteeg, M., F. Amadji, A. Eteka, A. Gogan and V. Koudokpon (1998) 'Farmers' adoptability of *mucuna* fallowing and agroforestry technologies in the coastal savanna of Benin', *Agricultural Systems*, vol 56, pp269–287

Wade, R. (1988) *Village Republics*, Cambridge, Cambridge University Press

Walker I. and J. Medina (2000) 'Agenda para la competividad y el desarrollo sostenible en el Siglo XXI' *Cambio Empresarial*, vol 11, no 8, Tegucigalpa, Honduras

Waters-Bayer, A., L. van Veldhuizen and C. Reij (2001) 'An encouraging beginning' in C. Reij, and A. Waters-Bayer (eds) *Farmer Innovation in Africa*, London, Earthscan

Watson, E. (2004) 'Agricultural intensification and social stratification: Konso in Ethiopia contrasted with Marakwet', in M. Widgren and J. Sutton (eds) *Islands of Intensive Agriculture in Eastern Africa*, London, The British Institute in Eastern Africa

Way, M. J. and H. F. van Emden (2000) 'Integrated pest management in practice — pathways towards successful application', *Crop Protection*, vol 19, pp81–103

Wedum, J., Y. Doumbia, B. Sanogo, G. Dicko, and O. Cissé (1996) 'Rehabilitating degraded land: *Zaï* in the Djenné Circle of Mali', in C. Reij, I. Scoones and C. Toulmin (eds) (1996) *Sustaining the Soil: Indigenous Soil and Water Conservation in Africa*, London, Earthscan

Whiteside, M. (2000) Ganyu *Labour in Malawi and its Implications for Livelihood Security Interventions – An Analysis of Recent Literature and Implications for Poverty Alleviation*, AgREN Paper 99, London, ODI

Widgren, M. and J. Sutton (eds) (2004) *Islands of Intensive Agriculture in Eastern Africa*, London, The British Institute in Eastern Africa

Wiersum, K. (1994) 'Farmer adoption of contour hedgerow intercropping: A case study from east Indonesia', *Agroforestry Systems*, vol 27, pp163–182

Wijeratne, M (1988) *Farmer, Extension and Research in Sri Lanka: An Empirical Study of the Agricultural Knowledge System with Special Reference to Matara District*, Wageningen, Wageningen Agricultural University

Wijeratne, M. (1993) 'Agricultural extension coverage under different management systems: status of small farmer', *Rohana*, vol 4, pp59–75

Wilken, G. C. (1987) *Good Farmers: Traditional Agricultural Resource Management in Mexico and Central America*, Berkeley, CA, University of California Press

Winarto, Y. T. (2002) 'From farmers to farmers, the seeds of empowerment', in M. Sakai (ed) *Beyond Jakarta: Regional Autonomy and Local Societies in Indonesia*, Adelaide, Crawford House

Winters, P., C. Crissman, P. Espinosa (2004) 'Inducing the adoption of conservation technologies: Lessons from the Ecuadorian Andes', *Environment and Development Economics*, vol 9, pp695–719

Winters, P., P. Espinosa and C. Crissman (1998) *Resource Management in the Ecuadorian Andes: An Evaluation of CARE's PROMUSTA Program*, Social Science Department Working Paper 1998–2, Lima, CIP

Woolcock, M. (2002) 'Social capital in theory and practice', in J. Isham, T. Kelly and S. Ramaswamy (eds) *Social Capital and Economic Development*, Cheltenham, UK, Edward Elgar

Woolcock, M. and D. Narayan (2000) 'Social capital: Implications for development theory, research and policy', *World Bank Research Observer*, vol 15, pp225–249

World Bank (2003) *Scaling-Up the Impact of Good Practices in Rural Development*, Agriculture and Rural Development Department, Report No 26031, Washington, DC, World Bank

Yunlong, C. and B. Smit (1994) 'Sustainability in agriculture: A general review', *Agriculture, Ecosystems and Environment*, vol 49, pp299–307

Zaal, F. and R. H. Oostendorp (2002) 'Explaining a miracle: Intensification and the transition towards sustainable small-scale agriculture in dryland Machakos and Kitui Districts, Kenya', *World Development*, vol 30, pp1271–1287

Zhu, Y., H. Chen, J. Fan, Y. Wang, Y. Li, J. Chen, J. Fan, S. Yang, L. Hu, H. Leung, T. Mew, P. Teng, Z. Wang and C. Mundt (2000) 'Genetic diversity and disease control in rice', *Nature*, vol 406, pp718–722

Index

ACORDE 96
Adégbidi, A. 22, 24
Africa 5–6, 14, 28–31 *passim*, 39, 47–8,
  51, 57, 75, 76, 82, 84, 86, 89, 91 *see
  also individual countries*
age factors 107, 112, 117
agriculture, sustainable 3–13, 44–5,
  206, 216
    LEISA 10
    LISA 5–6
Agrisystems (EA) Ltd 128
agro-ecology 8, 9, 36
agroforestry 14, 29–30, 58, 66, 72, 73,
  83, 85, 92, 98
Aiyar, S. 211
Ali, A. 37, 87
Ali, M. 79
alley cropping 27, 29–30, 33, 69, 75,
  76, 79, 91, 194, 216
Altieri, M. 8, 9, 14, 15, 33, 34
Amanor, K. S. 48, 58
Anderson, S. 22, 26–7, 37
Arrellanes, P. G. 98, 104, 109, 110,
  112, 120, 121
Ashman, D. 62
Asia 1, 7, 11, 14, 35, 38, 39, 41, 46–9
  passim, 51, 79 *see also individual
  countries*
Atampugre, N. 24, 73, 78, 88
Australia 54, 86

Baland, J.-M. 86
Barbier, E. 77
Barreto, H. 33, 76
Barrett, C. 9, 72, 80, 82, 83, 85, 92
Bassett, T. J. 47
beans 95, 120–1, 132, 134
Bebbington, A. 2, 44–5, 61, 62, 90
Belknap, J. 92
Benin 22, 24, 27, 76, 83, 85–6
Bentley, J. W. 37, 57
Berdegué, J. 70, 205
Besley, T. 54
Biggs, S. 40
Bindlish, V. 54
Binswanger, H. 211
biomass transfer 29, 75, 83

Birch-Thomsen, T. 82
Birkhaeuser, D. 54, 55
Blaikie, P. 49
Böhringer, A. 65
Borlaug, Norman 5–6
Boyd, C. 64, 75, 82
Branford, S. 90
Braun, A.R. 40
Bray, F. 46
Brazil 11, 26, 90, 92
    Friends of the Land clubs 26, 92
    Landless Movement 90
Brokensha, D.W. 89
brown plant hopper 171, 181
Brown, D. 72, 81, 92
Brown, J. S. 52
Brown, K. 76–7
Brown, L. D. 62
Bruin, G. 80, 92, 211
Brush, S. 84, 91
Bryceson, D. 49
Buckles, D. 22, 30, 80, 216
Budelman, A. 40
Buhler, W. 35, 79
Bunch, R. 24, 25, 40, 41, 76, 77, 81,
    87, 96–8 *passim*, 106, 110, 211
bunds 23–5 *passim*, 73, 75, 78, 81, 88
Burkina Faso 24, 26, 31, 59, 63, 64,
    71–3 *passim*, 78, 79
    Groupements Naam 88
Burton, M. 86, 91
Buttel, F. 4, 13–14
Butterworth, J. 11, 22, 28
Byerlee, D. 53

Cameron, L. 54–5
Campbell, B. 214–15
Campesino a Campesino movement 37,
  212
Canada 83, 86
Carlson, G.A. 83
Carson, Rachel 1
Carter, J. 30, 69, 79
Carter, S. 71, 82
Cary, J. 86
Case, A. 54
Cassman, K. G. 12, 56

cattle 76, 132, 135, 152
Central America 22, 24, 30–1, 40, 77, 81, 87, 211, 212, 216 *see also individual countries*
Central Bank of Sri Lanka 165
CGIAR 217
Chaker, M. 75
Chambers, R. 2, 40
Charles, D. 53
China 6, 11, 72, 92
Choi, H. 52
CIP-UPWARD 40
coffee 75, 95, 101, 102, 110, 126, 132
Coleman, J.S. 59, 65
Collinson, M. 2, 47
Colombia 46
colonial era 7–8, 89, 126
composting 11, 27–9 *passim*, 71
Conklin, H.C. 57, 60
Conley, T. 63–4
contour planting/ploughing 23, 24, 149
Conway, G. 2, 3
Cooper, P. 25, 29, 30
COSECHA 96
Côte d'Ivoire 47
cotton 36, 47, 75, 126, 134
Coughenour, C. 52
Coutts, J. 39
cover crops 20, 23, 26–7, 33, 71, 75, 76, 80–86 *passim*, 91, 98, 188, 189, 198, 216
Cowan, R. 84, 90, 92
Cramb, R. A. 69, 75, 76, 84, 90, 213
credit 62, 117
Crowley, E. 71, 82

Daberkow, S. 84
David, C. 47
de Groot, J. 9, 84, 90–1
de Jager, A. 33, 75
de Janvry, A. 205
de Renzio, P. 60
deforestation 87
Defour, T. 40
degradation, environmental 2, 86–7 resource 12, 23, 126
demand, farmer 214–15, 217
Dembélé, I. 79
Desmodium 34, 149
development, agricultural 1–42 *passim*, 204–6, 208, 213
Díaz, A. E. 96
Dilts, R. 88

discount rates 3, 13, 85, 87
ditches, infiltration/retention 23, 131, 139, 140
Doberman, A. 32
Dominican Republic 85
donors 89, 207–10 *passim*, 212, 214, 217
Douglass, G. 3
Douthwaite, B. 76, 91
drains 23, 81, 129, 131, 139
drought 88, 114, 163
Duguid, P. 52
Dvořák, K.A. 69

economic environment 74–6
Ecuador 73, 76, 79, 81, 83, 90, 92
education 19, 43, 51, 53, 55–7 *passim*, 60, 65, 67, 69, 83–4, 87, 94, 168, 179–80, 197, 214, 216 *see also* Honduras
Ekboir, J. 26
El Salvador 76
Ellis, E. 6
Ellis, F. 49, 51
Elvin, M. 6
employment 48–9, 73, 196 non-farm 49–51, 81–3 *passim*, 93
environmental concerns 1–3, 85–7, 200
equity issues 66, 69–74
erosion 23, 25, 26, 38, 72, 77, 95, 97, 98, 126, 150, 193, 200, 215
Escalada, M. 84
Escobar, G. 70, 205
Ethiopia 47, 48, 64, 88–9
Europe 4, 79
Evenson, R. 54
experimentation, farmer 15–17 *passim*, 36, 40, 57, 63–4, 80, 84, 97, 113, 149–50, 174, 185, 197, 198, 200–1
extension 2, 14, 37–42 *passim*, 47, 54, 55, 58–9, 73, 84–5, 87, 89, 97, 116, 128–9, 151, 154, 161, 166–8, 184, 214–16 *passim*
extensionists, farmer 17, 38, 41, 91

Fakih, A. M. 73, 88, 89
Falck, M. 96
fallowing 6, 15, 30, 42, 58, 71, 72, 83
FAO 20, 39, 161, 175
Farrington, J. 2, 211
Feder, G. 43, 44, 54–6 *passim*
Fernandez-Cornejo, J. 52
fertilizer 1, 2, 5–9 *passim*, 11, 27, 31–2, 46, 56, 76, 79, 107–9 *passim*, 115,

117, 121–3 *passim*, 151, 166,
173, 174, 180, 198, 199
organic 6, 27–8, 98, 104, 115, 116,
198
FFS 19, 20, 35, 38–42 *passim*, 73, 80,
91–3 *passim*, 197, 208, 210–11, 216
see also Sri Lanka
Fiji 25
Fine, B. 60
Fischer, A. J. 54
Fleischer, G. 210
food aid 89
food for work 78, 137
Foster, A. D. 55, 63
Fox, H. 65
Francis, C. 33
Francis, P. 75
Franzel, S. 14, 72, 211
Freebairn, D. 70, 204, 207, 208
fungicides 11, 33, 115, 166, 171, 183,
199
Fujisaka, S. 84

Gambetta, D. 60
Gautam, M. 54
Geertz, C. 7
Ghana 48, 55–6, 58, 63
Giller, K. 28
Global 2000 37, 47, 89
Gonsalves, J. F. 211, 212
Granovetter, M. S. 61
Graves, A. 13, 28–30 *passim*, 212
Green Revolution 1–3 *passim*, 5, 7, 35,
38, 46–9 *passim*, 53, 55, 70–2 *passim*,
88, 89, 204, 205, 207, 208, 216
Greenpeace 5
Grieshop, J. 80, 84
Griffin, K. 1, 205
Grindle, M. S. 50
Grootaert, C. 59, 60, 63, 64
groundnuts 132, 134
groups/group formation 17, 38–9,
60–7, 88, 90, 94, 114, 151–2, 168,
197–8, 202–3, 207, 209, 215
Guatemala 9, 22, 25, 77, 84, 90–1, 96,
211
Gugerty, M. K. 66
Gunby, P. 84, 90, 92
Gündel, S. 211

Haggblade, S. 5–6, 32, 75, 81, 85
Haiti 75
Harding, D. 127
Harrison, E. 66

Harrison, P. 30
Harriss, J. 60, 61
Harriss-White, B. 51
Hassan (Hassane), A. 31–2, 71, 81
Hayami, Y. 46
Hazell, P. B. R. 48, 70
hedgerows 23, 25, 69, 72, 75, 81, 84,
92
Heinrich, J. 54
Hellin, J. 25, 37, 77, 97, 111, 112, 116
Heong, K. 37, 84
herbicides 5, 11, 26, 33, 49, 84, 115,
123, 166, 171, 175, 198
Hesse-Rodríguez, M. 98
Hine, R. 208
Hirschmann, A. 61
Hogg, R. 78
Holmberg, G. 126
Holt-Giménez, E. 37, 41, 87, 121, 212
Honduras 20, 22, 25, 30, 57, 64, 71,
75–7 *passim*, 80, 84, 87, 95–124,
192–203 *passim*, 211, 216
adoption of LEIT 96, 103–11, 115
capital, human/social 96, 100–1,
107, 111–15, 117
Comisajul Cooperative 111–12
education 107, 110, 112, 117
GRMs 98, 104–7, 109, 111, 114–16
IRT 98, 104–17 *passim*, 122, 123,
188, 189, 192–5 *passim*, 198, 201,
202
labour 110–12, 117, 122, 123, 192,
195–6
Lempira Sur project 97
Howard-Borjas, P. 79
human capital 15–20 *passim*, 55–9, 67,
69, 83–8, 208 *see also* Honduras;
Kenya; Sri Lanka
Humphries, S. 64, 113

ICRAF 14
incentives 19, 47, 51, 69, 70, 74–8, 82,
83, 94, 129–30
income 48, 70, 82, 110, 132, 134, 135,
141, 146, 147, 167, 192, 195
non-farm 49–52, 70, 81–3 *passim*,
192, 196
India 25, 29, 48–9, 51, 55, 56, 60, 62–4
passim, 70, 73, 81, 83, 85, 87, 211
Indonesia 7, 35, 39, 72–5 *passim*, 77,
82, 88, 89, 91, 92, 211
industrialized countries 1, 2, 4, 8, 21,
53, 83, 86 *see also individual entries*
information 13, 15–19 *passim*, 22, 42,

52–5, 63, 67, 148–9, 179–81, 215–16
    farmer transmission 16, 17, 20, 54,
        63–4, 90–1, 93, 116, 148–53,
        175–81 *passim*, 184, 185, 201–2,
        204
    flow 90–3, 113–14, 117, 153,
        175–9, 184
innovation, farmer 15–20, 36–41
    passim, 46, 56–7, 61, 63–4, 67, 84,
    97, 113, 150, 153, 174, 185, 215–17
    passim
insecticides 1, 11, 33–6 *passim*, 91, 123
    see also Sri Lanka
intercropping 9, 11, 22, 25, 29–30,
    32–4 *passim*, 72
IPM *see* pest control
Iraq 6
irrigation 2, 9, 46, 48, 50, 64, 65, 71,
    95, 107, 108, 162
Italy 60

Jansen, H. G. 95, 109, 111, 112, 121
Jansen, K. 79
Japan 46
Jeger, M. 35, 36, 69
Johnson, D. 23
Jones, K. 91, 179
Jones, S. 53

Kaboré, D. 31, 37, 71
Kaboré, V. 79
Keeley, J. 88–9
Kelly, V. 72
Kenmore, P. 35, 39
Kenya 20, 23, 34, 48, 50, 51, 66, 71–3
    passim, 75, 78, 82, 83, 85, 89, 91, 92,
    125–59, 188–203 *passim*
    adoption of LEIT 138–48, 153
    capital, human/social 125, 150–4
        passim
    Catchment Approach 20, 126–31,
        137–48, 150–4, 188
    catchment committees 129–30,
        137–8, 194, 197, 203
    fanya juu 23, 130–1, 139, 142, 149
    KICOBADE project 152
    NALEP 127
    soil/water conservation 20, 25,
        125–59 *passim*, 188, 189, 192,
        195, 200, 201
Kerr, J. 49, 73, 81, 83
Khan, Z. 34
Kimani, M. 91
knowledge, farmers'/ITK 2, 4, 15,

36–7, 56–8, 67, 150–1, 168, 180,
    196–8
Kolavalli, S. 49
Konde, A. 64
Korte, C. 72, 81, 92, 93
Kremer, M. 66
Krishna, A. 60, 62
Kuyvenhoven, A. 15

labour 5, 7, 15, 16, 19, 22–4, 27–31
    passim, 41, 42, 45–52, 59, 66–7,
    78–83, 93, 174, 182, 194–6 *see also*
    Honduras
    communal 47–8
    costs 46, 51, 79–81, 110, 111
    ganyu 50
    hired 48, 67, 81, 166, 194–6
    off-farm 49–51, 81–2, 194, 195
    timing 41, 47, 78–9, 81, 82, 194
land 6, 23, 50, 71, 76, 95, 107, 121
    reform 90
    tenure 74–5, 117
Landcare Movement 39, 213
Landers, J.N. 26, 92
Latin America 14, 26, 40, 71, 75 *see
    also individual countries*
Leach, M. 86
Leaf, M.J. 48
learning 16, 38–40, 42, 51, 80, 82, 94,
    111, 215–16
Lee, D. 71, 75, 76, 82, 84, 121, 216
LEISA Magazine 40
LEIT 10–52, 53, 59, 68–94, 187–217
    abandonment 105–6, 111, 188
    adaptation 16–19 *passim*, 22, 37,
        56, 125, 198, 200–1
    adoption 14, 16, 18–20 *passim*,
        43–5, 51–2, 54–5, 68, 70–79
        passim, 83, 93, 94, 193, 207 *see
        also* Honduras; Kenya; Sri Lanka
    assessment 13–17, 19, 68–94, 206–9
    costs 13, 210–12
    definition 11–13
    examples 21–42
    utilization 16–19, 68–94, 187–94,
        199, 203–4 *see also* Honduras;
        Kenya; Sri Lanka
Lele, U. 217
Lewis, D. 89
Lewis, L. 23
Lightfoot, C. 33
Lin, N. 59
Lipton, M. 1, 48, 70, 72
Little, P. D. 89

livelihood diversification 19, 44–5, 182–3, 205–6
Lobo, C. 211
Lockeretz, W. 44, 54, 86
Loevinsohn, M. 19, 38, 207, 210
Long, N. 64
Longhurst, R. 1, 48, 70, 72
Longley, Catherine 125–59
Low, A. 5, 49
Lumley, S. 72, 85
Lutz, E. 75, 111

Madagascar 12, 32, 72, 78, 82, 83, 85, 92
Madeley, J. 5
maize 34, 49, 55–6, 71, 75, 76, 104, 105, 120–3 *passim*, 132, 135, 201
Malawi 50
Mali 31
Malley, Z. 31
Mango, Caleb 125–59
Mango, Nelson 125–59
Mansuri, G. 207
manuring 11, 15, 21, 27–30 *passim,* 72, 76, 150
    green 22, 26, 30, 79, 174, 177, 180, 182, 185, 216 *see also* Honduras
Manyong, V. M. 37, 83, 86
marijuana 75
markets 49, 71, 75, 76, 93, 104, 110–12 *passim*, 117
Marsden, T. 5
Masdewel, M. 26, 72
Mayoux, L. 66
McBride, W. 52, 84
McDonald, M. 76–7
Mdoe, N. 49
Mearns, R. 86
mechanization 2, 46, 48, 49
Medina, J. 95
Meerman, F. 80, 92, 211
Meinzen-Dick, R. 19
Mercado, A. R. 24–5, 39, 76, 213
Mexican sunflower 29
Mexico 12, 50, 57, 80, 81, 216
Miclat-Teves, A. 89
migration 50–1, 79, 81, 82
Miller, D. 211, 213
Mitchell, J. C. 61
Moi, Daniel arap 127–8
Morocco 75
Morris, M. L. 55–6
Morse, S. 35, 79
Mortimore, M. 214–15

Moser, C. 72, 80, 82, 83, 85, 92
mulching 23, 25, 26, 33, 71–2, 76
Murray, G. F. 77, 85
Murton, J. 50
Murwira, K. 57

Nagy, J. 24
Napier grass 25, 34, 129, 130, 140–2 *passim*, 193
Narayan, D. 60, 62–3
Naylor, R. 33, 49
Ndakidemi, P. 28
Neill, S. 71, 75, 76, 82, 84, 216
Nepal 5, 49
Netting, R. McC. 64
networks 17, 61–4 *passim*, 67, 90–3, 117, 154, 203
NGOs 2, 5, 14, 29, 50, 85, 87–9 *passim*, 95, 148, 211, 214, 217
Nicaragua 37
Niger 31, 71, 72, 81
Nigeria 47, 75
Nindo, Wilson 125–59
nitrogen fixation 4, 27, 84
Noordin, Q. 92
NRC 4, 25
Nugaliyadde, L. 166
nutrient cycling 3, 4, 8, 27, 84

Okali, C. 57, 63, 64, 201
Okoth, J. 39
onion maggot 34
Oostendorp, R. H. 75
organizations, community 15–16, 20, 60–6, 88, 91–3, 111–12, 114–15, 151–4 *passim*, 168, 202–3, 207, 213, 214, 217
Orr, A. 36, 37, 75, 84
Orton, L. 5
Ostrom, E. 60
Otsuka, K. 47
Ouedraogo, M. 79
Owens, M. 92

Palladino, P. 35
Parrott, N. 5
participation, farmer 2, 16, 38, 40–1, 54, 190–4
Pender, J. 70, 81, 83, 85, 97, 110, 111
Perreault, T. 61
Peru 50
pest control 4, 19, 20, 33–6, 57, 87, 90, 92, 149, 161, 169–72, 177–81, 183–5, 195, 201

IPM 11, 12, 35–7 *passim*, 39, 52, 69, 71, 73, 75, 79–84, 88–92 *passim*, 160, 175, 184, 192, 194, 197, 202, 208, 210–11, 215
IPPM 39
pesticides 1, 2, 7, 8, 11, 33–6, 58, 92, 115, 122, 200, 210
Philippines 24–5, 39, 57, 69, 72, 74–6 *passim*, 81, 84, 85, 89–93 *passim*, 211, 213
Pingali, P.L. 34, 83
Piters, B. 28
pits, planting 31–2, 37, 71, 75, 79, 81
Piyadasa, V. Hiroshini 160–86
Place, F. 14, 66, 73–5 *passim*, 83, 85, 129, 150, 212
Platteau, J.-P. 86
plough pans 2, 32
political action 61, 62, 88–90, 94
Ponnamperuma, F.N. 181
Pontius, J. 39
poverty 1, 6, 12, 14–16 *passim*, 69–74, 110, 117, 196, 205
    reduction 71, 74, 196
Pretty, J. 2–4 *passim*, 7, 9, 12–15 *passim*, 34, 38, 44, 65, 71, 79, 127, 129, 208, 212
Price, L. 91
Pritchett, L. 62–3
projects 187, 190, 207–14, 217
Putnam, R. D. 60, 61, 64, 65

Quizon, J. 41, 91, 211

Ramasamy, C. 48, 70
Rao, V. 207
Reardon, T. 50–1, 56
Reij, C. 22, 31, 37, 57, 71, 84
Reijntjes, C. 10. 12
Reinhardt, N. 46
remittances 135
Renkow, M. 70
research 2, 40, 70, 84, 214–17 *passim*
    FSR 2, 40
    PLAR 40–1
rice 1, 7, 11, 12, 22, 35–6, 39, 46–9 *passim*, 71, 79, 82, 160–86 *passim*, 188
    SRI 32, 42, 72, 78, 80, 82, 83, 85, 92
    straw incorporation 173, 176–85 *passim*, 188, 195, 199
Richards, Michael 95–124
Ridgeley, A.-M. 84, 91
ridging, contour/tied 24, 78–9

Roberts, K, 39
Rocha, J. 90
Rogers, Everett 4–5, 44, 61, 205
Rola, A. C. 34, 80, 91
Röling, N. 35, 73
Rosenzweig, M. R. 55, 63
rotation, crop 9, 15, 34, 95
Ruben, R. 70, 111, 115
Ruthenberg, H. 45–6
Ruttan, V.W. 46, 88
Rwanda 23

Sadoulet, E. 205
Sahel 24, 31, 37, 79
Sain, G. 33, 76
Sanders, R. 72, 92
Saupe, W. E. 92
Sayer, J. 214–15
scaling up 14, 20, 211–13
Scarborough, V. 41
Scherr, S. 97, 111
Scholte, E. 214
Schrader, K. 25, 37, 77, 97, 111, 113, 116
Schultz, T. W. 55
Scoones, I. 82, 84, 88–9
seed 12, 70, 71
Sen, A. 55
Settle, W. 36
Seyoum, E. 47
Shapiro, Robert 53
Shiva, V. 33
Sida 127
Sierra Leone 22
Simpson, B. 92
sisal 134, 142
skills 4, 9, 12, 15–16, 19, 51, 52, 55–9, 69, 79–84 *passim*, 94, 194
Slade, R. 54, 56
Slaymaker, T. 64, 75, 82
Slingerland, M. 26, 72
Smale, M. 88
Smil, V. 6, 7
Smit, B. 3
Smith, K. R. 52
Smucker, G. 75
social capital 15–20 *passim*, 38, 44, 58–67, 88, 90–3, 202–3, 208, 212 *see also* Honduras; Kenya; Sri Lanka
    definition of 59–62
soil, conservation 9, 19, 20, 22–7, 37, 48, 57, 64, 69, 72–85, 88–90 *passim*, 93. 96, 98, 112, 116, 188, 213
    see also Kenya

fertility 2, 7, 9, 19, 20, 27–31, 40,
46, 75, 85–7 *passim*, 96, 98, 150,
173–4, 177–80, 183, 185, 188,
199, 202
regeneration 4, 7, 72, 77, 150, 188
see also Honduras
Soule, M.J. 80, 81
SRI *see* rice
Sri Lanka 20, 91, 160–86, 188, 189,
192–203 *passim*, 211
adoption, of LEIT 179–83
capital, human/social 160, 168,
175, 185
FFS 161–3, 166–79, 183–5, 188,
190–1, 197, 198, 201
insecticide use 161, 166, 169,
171–80, 182–5, 188, 189, 195,
197, 199–200
stem borers 34
Stocking, M. 37, 69
Stoll, S. 72
Stone, G. D. 47
stone lines 141, 142, 150
Stoop, W. 32, 80
striga 34, 149
strips, unploughed 130, 139–50 *passim*,
188, 194
vegetative 20, 24–5, 75–6, 81, 130,
139–50 *passim*, 153, 188, 193,
194
Suazo, Laura 95–124
subsidies 1, 2, 6, 185
sugar 7, 132
Sumberg, J. 57, 63, 64, 201
sustainable livelihoods 44–5, 59–66
Sutton, J. 22
Swaziland 49

Tanzania 28, 31, 49, 62–3, 82
tea 132, 165
Tembo, G. 32, 75, 81, 85
terracing 20, 21, 23–4, 48, 71, 75, 81,
89, 92, 110–11, 140, 150, 188
fanya juu 23, 130–1, 139, 142, 149
Thomas, D. B. 126, 130, 131
Tiffen, M. 23, 82, 89, 126, 140
tillage, conservation 11, 12, 26–7, 32,
42, 71, 76, 78, 84, 90, 92
in-row 20, 24 *see also* Honduras
reduction/elimination 11, 23, 26–7,
33, 52
Tisdell, C. 3
tomatoes 80, 84, 111, 115, 122, 123
trade 50, 135, 146, 147, 182, 192, 196

training 97, 129, 197
Training and Visit 54, 126, 161
transport 79, 112
Traoré, N. 83, 86
tree nurseries 65–6, 151–2
trenches 148–9
Tripp, Robert 1–94, 160–218
trust 60, 62, 63, 114–15

Udry, C. 63–4
UK 65, 86, 91
unemployment 49
United Fruit Co. 22
Uphoff, N. 8, 12, 15, 78
US 4–5, 14, 22, 26, 34, 52, 60, 72, 80,
83–4, 86, 90–2 *passim*
Uvin, P. 211, 213

Vaessen, J. 9, 84, 90–1
van Bastelaer, T. 60
van den Berg, H. 166, 173, 175, 208,
211
van Emden, H. F. 36, 69, 216
van de Fliert, E. 35, 73
van der Ploeg, J. 46, 50
vegetables 36, 75, 77, 95, 101, 104–10
passim, 115, 117, 134, 174, 192, 193,
200
velvet bean 22, 27, 30–1, 37, 76, 80,
83, 85–6, 98, 216
Veron, R. 87
Versteeg, M. 27, 85–6
vetiver grass 25, 216
Vietnam 84
visits, cross-/exchange 37, 40

Wade, R. 64
wages 48–9, 122
Walker, I. 95
Wang, S. 6
Ward, H. 44, 65
water conservation 9, 19, 20, 22–7, 37,
48, 57, 64, 75–8 *passim*, 82, 88, 96,
98, 213 *see also* Kenya
watershed development 73
Waters-Bayer, A. 37, 57, 84
Watson, E. 48
Way, M. J. 36, 69, 216
Wedum, J. 31
weed control 22, 23, 26, 27, 33–4, 47,
49, 58, 91, 98, 166, 171
Wernick, S. 86
wheat 1, 12, 46–7, 54
Whiteside, M. 50

Widgren, M. 22
Wiersum, K. 72, 74, 75
Wijeratne, Mahinda 160–86
Wilken, G.C. 22, 57, 58
Wilkinson, R. 86
Winarto, Y. T. 82, 92
Winters, P. 73, 76, 79, 81, 83, 92
women 74, 78, 80, 97, 152
Woolcock, M. 59, 60, 62
work parties 48, 64
World Bank 25, 54, 61–3 *passim*, 211, 213

World Neighbors 95, 96

yields 2, 13, 15, 32, 46, 115, 122, 123, 165, 181, 189
Yunlong, C. 3

Zaal, F. 75
Zambia 32, 66, 72, 75, 80–1, 85
Zanzibar 80, 92, 210–11
Zhu, Y. 11
Zimbabwe 57